SpringerBriefs in Environmental Science

SpringerBriefs in Environmental Science present concise summaries of cutting-edge research and practical applications across a wide spectrum of environmental fields, with fast turnaround time to publication. Featuring compact volumes of 50 to 125 pages, the series covers a range of content from professional to academic. Monographs of new material are considered for the SpringerBriefs in Environmental Science series.

Typical topics might include: a timely report of state-of-the-art analytical techniques, a bridge between new research results, as published in journal articles and a contextual literature review, a snapshot of a hot or emerging topic, an in-depth case study or technical example, a presentation of core concepts that students must understand in order to make independent contributions, best practices or protocols to be followed, a series of short case studies/debates highlighting a specific angle.

SpringerBriefs in Environmental Science allow authors to present their ideas and readers to absorb them with minimal time investment. Both solicited and unsolicited manuscripts are considered for publication.

More information about this series at http://www.springer.com/series/8868

Elise Machline · David Pearlmutter
Moshe Schwartz · Pierre Pech

Green Neighbourhoods and Eco-gentrification

A Tale of Two Countries

 Springer

Elise Machline
Ben Gurion University of the Negev
Blaustein Institutes for Desert Research
Midreshet Ben Gurion, Israel

Urban and Mobility Department
Luxembourg Institute of Socio-economic
Research
Esch sur Alzette, Luxembourg

Moshe Schwartz
Bona Terra Department of Man
in the Desert
Ben-Gurion University of the Negev
Sde Boker, Israel

David Pearlmutter
Bona Terra Department of Man
in the Desert
Ben-Gurion University of the Negev
Sde Boker, Israel

Pierre Pech
Department of Geograph
Pantheon-Sorbonne
Paris, France

ISSN 2191-5547   ISSN 2191-5555   (electronic)
SpringerBriefs in Environmental Science
ISBN 978-3-030-38035-9   ISBN 978-3-030-38036-6   (eBook)
https://doi.org/10.1007/978-3-030-38036-6

This Springer imprint is published by the registered company Springer Nature Switzerland AG
The registered company address is: Gewerbestrasse 11, 6330 Cham, Switzerland

Preface

With momentum building for an expansion of architectural and urban design practices that respond to the environmental challenges of our time, it is worth considering the economic and social implications of what has come to be known around the world as "green" building.

In this monograph, we bring together a series of studies that delve into the details of green building practices in France and Israel and that tell a tale of two countries that deviates considerably from what first impressions might suggest. In-depth data analysis, interviews with stakeholders, and on-the-ground documentation are used to paint a portrait of green neighborhoods in both large and small cities and to shed light on the diversity of outcomes and the intricate web of interests leading to each one.

- We begin by summarizing the development of "green" building in both Israel and France, shedding light on both countries' specificities: In Israel, there is a lack of any national legislation fostering "green" building practices and, at the same time, an acute shortage of affordable housing, while in France, the prevalence of social welfare policies has produced legislation, officially promoting "green" and affordable housing.
- The second chapter points out that the target population of "green" building projects is usually the middle to upper classes and such targeting ultimately perpetuates socio-spatial inequality as well as ecological vulnerability for the poor and other socially marginal groups. We compared policy contexts and "green" building instruments in France and Israel and considered whether affordable housing and social diversity are part of green building policy and implementation. We also inquire whether green buildings foster gentrification, either inadvertently or intentionally.
- In the third chapter, we have addressed the impact of the Israeli green building certification on real estate prices. Our conclusion is that in centrally located and economically strong municipalities, this involves green certification, while in peripheral locations, such certification is not implemented, and the "green" label is mainly used to attract local residents who can afford a housing upgrade.

- We then describe how French policy promotes social diversity and the construction of "green" public social housing in the growing series of urban "eco-districts." While there is an ostensible effort to build housing that is both "green" and affordable, it turns out that in affluent and average municipalities, the share of "green" social public housing actually available to low-income groups is minimal, since most social housing is ultimately allocated to higher-income groups.
- In our concluding discussion, we explain that "green" building has yet to prove itself as a solution for the masses. The sale price of an apartment in a certified green building is significantly higher than what would be justified by either the additional construction costs required to build it or the energy and water-saving potential that can be realized by using it. The tale of two countries presented here suggests that neither the mechanisms of the market nor the proclamations of a welfare state can easily overcome this dilemma. What is needed is a new type of thinking, which can only emerge once the concept of "value" reflects not only the realities of a free-market economy but also those of a planet which turns out to be distinctly limited in its resources.

Sde Boker, Israel and Esch Sur Alzette, Luxembourg   Elise Machline
Sde Boker, Israel   David Pearlmutter
Sde Boker, Israel   Moshe Schwartz
Paris, France   Pierre Pech

Acknowledgments

The authors would like to thank the interviewees and experts who contributed to this study by sharing their precious experience openly and voluntarily. Our thanks to Francine Davidai, head of the Northern Tel Aviv Planning Unit; Cedissia About de Chastenet, head of the Housing Authority of Paris municipality; Bruno Bessis, head of Sustainable Planning Unit of the French Ministry of Housing; and officials from the municipalities of Bretigny sur Orge, Reims, Yavneh, and Dimona. We are grateful to the numerous eco-district residents who responded to our surveys.

Contents

Introduction 1

The Green Building Agenda 3

The Socio-Economic Impacts of 'Green' Building in Israel: Green
Building as an Urban Branding Tool 21

Green Building in Europe 61

The French Case Studies 69

Discussion-Conclusion 87

Appendix 1 95

Appendix 2: Post Occupancy Survey in Yavneh's 'Green'
Neighborhood 97

References 99

Index 107

Introduction

Under the banner of "green building", policy instruments have been developed worldwide, to reduce the energy demand and overall environmental impact of buildings. Using green rating systems and other tools, these policies have often improved the quality of the building stock – but at high cost to consumers. Studies of the added construction costs involved in achieving green certification have suggested that green apartments need not be significantly more expensive than non-green alternatives – but evidence is accumulating that the prices charged to buyers and renters of such properties are higher indeed.

In this context, the "green premium" of real estate has become an important issue in developed countries. On the one hand, reducing operating expenses through green measures is widely seen as a way to increase the long-term viability of development, as tenants benefit from lower utility bills and enjoy indirect economic benefits, such as improved indoor air quality and long-term occupant health. On the other hand, however, higher housing costs tend to perpetuate inequality, as well as ecological vulnerability for the poor and other socially marginal groups. A significant green premium can defeat the goals of sustainable development – which include socio-economic and as well as environmental dimensions (WCED 1987). There is irony in this situation, considering that while living in green housing may reduce energy related expenses, those who might benefit the most from these savings, cannot afford the initial cost of accessing them.

Contrasting approaches to green building are illustrated by France and Israel, which have both promoted the certification of green residential construction since the early 2000s. While a broad consensus has been built in both countries around the environmental benefits of green building, the same cannot be said about the need to make it affordable to a cross-section of the population. In Israel, as in many other countries, "green" real estate is primarily marketed to the upper middle-class – whereas in France, where social welfare policies are deeply ingrained, providing subsidized green housing is an official goal.

© The Author(s), under exclusive license to Springer Nature Switzerland AG 2020   1
E. Machline et al., *Green Neighbourhoods and Eco-gentrification*,
SpringerBriefs in Environmental Science,
https://doi.org/10.1007/978-3-030-38036-6_1

With momentum building in both countries to expand green building, it is worth considering its economic and social implications. In recent years Israel has faced an acute shortage of affordable housing which has led to social unrest, and a major decision by the country's largest municipalities to require the implementation of the voluntary Green Building Standard in all new construction, could exacerbate the situation. France, for its part, has embarked on a large-scale initiative to build *eco-quartiers* – green residential districts in which measures enhancing socio-economic diversity are mandated alongside environmental performance criteria. Might such measures be relevant for Israel?

In this monograph, we bring together studies delving into the details of French and Israeli green building practices and tell a tale of two countries. We also describe the social aspects of 'green' building in the eco-districts of Stockholm and Copenhagen. The Scandinavian countries have a social housing policy, but 'green' building is no part of it. In-depth data analysis, interviews with stakeholders, and on-the-ground documentation are used to paint a portrait of green neighborhoods in both large and small cities, shedding light on the diversity of interest constellations and resulting outcomes.

More often than not, green initiatives are used to attract upper middle-class dwellers to previously poor neighborhoods –displacing the original residents through "green" gentrification. However, the form and scope of this eco-gentrification vary widely from one city or neighborhood to the next, depending on the <u>political,</u> administrative and, economic contexts.

In the Israeli cases, these dynamics reflect the increasing dominance of the private sector in residential building, following a decades-long shrinking of the welfare state. In the densely populated urban core of the country, exorbitant real estate prices mean that developers can profitably exploit the Israel Green Building Standard as a marketing tool – while in the economically depressed cities of the periphery, this is not the case and green certification is not sought.

The French way is to mandate the inclusion of subsidized housing within its *eco-quartiers*, with the declared aim of promoting a diverse 'social mix'. In Paris, however, most "public" housing is in fact intended for the middle class, and the eco-districts have primarily been located in areas where the local population is being priced out of the market – effectively forcing it out of Paris altogether. Moreover, in the one French case study which documents the establishment of green housing for the poor, social diversity turned out to be unattainable due to a lack of interest among those who can afford to live elsewhere.

In sum, this monograph brings together for the first time the evidence needed to answer a crucial question: If 'green' building does offer individual as well as societal benefits, can it be affordable to those who need it the most?

The Green Building Agenda

Over the last two decades, greenhouse gas abatement for climate protection has become a major goal in developed countries and increasing attention has been focused on how to make buildings ('whose ongoing operation consume about 40% of all energy') more efficient. "Green building" refers to "the use of environmentally preferable practices and materials in the design, location, construction, and operation of buildings. It applies to both renovation or retrofitting of buildings and construction of new ones, residential or commercial, public or private" (Commission for Environmental Cooperation 2008).

In addition to a building's design and construction (which directly affects the uses of energy, water, and materials), green building deals with environmental issues, ranging from ongoing building operation, to urban planning for reduced reliance on private cars. The introduction of neighborhood-level green building standards has further extended the scope of sustainability topics and stakeholders, including municipal authorities, whose role is on the increase.

To promote 'green' building implementation, broad spectra of policy instruments and programs have been enacted worldwide by governments and other decision-makers. In the literature on such policy tools, a distinction is generally made between regulatory, economic and informative/educational instruments (Vine et al. 2003).

Prominent among these are 'green' building rating systems and energy efficiency standards for buildings, developed around the world. The British rating system BREEAM (Building Research Establishment Environmental Assessment Method), established in 1990, later served as basis for the American LEED (Leadership in Energy and Environmental Design) and the Australian Energy Star. These rating systems assign credits to building projects submitted for certification in categories such as energy efficiency, water use efficiency, sustainable site selection, materials and resource use, and indoor environmental quality. In recent years, a few organizations have gone beyond building-level environmental assessments, initiating

Table 1 Main benchmarks of LEED and Breeam certifications

LEDD	BREEAM
Location and transportation	Management
Sustainable sites	Health and wellbeing
Water efficiency	Energy
Energy & atmosphere	Transport
Material & resources	Water
Indoor environmental quality	Materials
Innovation	Waste
	Land use and ecology
	Pollution

Source: http://www.breeam.org/page.jsp?id=369; USGBC (2011)

additional schemes relating to the neighborhood scale. For example, "BREEAM Communities" and "LEED-ND" (Neighborhood Design) were launched in 2009 (Hamedani and Huber 2012) (Table 1).

The Israeli Governance Regime

Israel is a centralized state. The land use planning activities of local governments are subordinate to district and national committees, controlled by the central government. Thus, municipalities have little involvement in public housing (Razin 2004). This centralization is expressed in the share of local government within overall government expenditure, which is low in Israel compared to most developed countries (International Monetary Fund 2000), sometimes attributed to the inordinately large proportion of defense expenditures incurred at the national level.

Substantial political decentralization has occurred since the late 1970s, and local government has gained strength. It has expanded into new activities, particularly economic development and welfare services, partly because of the governance vacuum created by the retreat of the central state (Ben-Elia 1998). Persisting deficits are a critical issue in relations between local and central government (Dery and Schwartz-Milner 1994). Local authorities complain about inadequate government transfers, coupled with government decisions and legislation that imply increased costs or reduced revenues (Razin 2004). Wealthier municipalities are less dependent on the State, which has more control over poor localities.

Israeli local authorities enjoy substantial autonomy coupled with insufficient fiscal accountability. According to Beeri and Navot (2013), 'since the early 1990s, there has been a dramatic rise in the use of the term corruption linked to lots attribution for housing construction'. The contracts and tenders that are an inevitable part of governing and that, by their nature, involve transferring public assets and resources to private hands and are source of corruption (Jain 2001). This is even more obvious for authorities controlling greater assets and resources, either because they are financially strong or for other reasons-such as their size-(Karahan et al. 2006). Administratively, Israel is divided into six districts: Central District; Haifa

District; Jerusalem District; Northern District; Southern District; and Tel Aviv District. The districts are subdivided into 15 sub-districts and 50 natural regions (see map in annex). District administration is coordinated by the Ministry of the Interior.

There are three forms of municipal government in Israel: city councils, local councils, and regional councils. City councils govern cities (20,000 residents and above), local councils – small municipalities, and regional councils – groups of rural settlements. The municipalities are mandated to provide some public services such as garbage removal, veterinary services, urban planning and zoning, water distribution, emergency services, education (partly) and culture. Local governments are run by elected governing councils chaired by elected mayors.

In Israel power is heavily concentrated in the central government. The small size of the country, the ways in which the British mandate has limited municipal power to maintain political control, which the state of Israel's ruling elites have maintained for partisan reasons and the development of the country through an ideologically motivated effort, have all contributed to the concentration of power. Local government in Israel is often viewed as the weakest link in the state's political system. From a power perspective, local governments are indeed subordinate to government and party centers.

Local governments play an important role in 'green' building development. Thus, the first 'green' neighborhood of Israel, in Kfar Saba, was developed by the Sharon region environmental unit, connected administratively to local authorities but advised by the Ministry of Environmental Protection. In 2009, the municipality of Ra'anana (in the central region) was the first to adopt the SI 5281 standard as mandatory (Goulden 2016). More importantly, in 2008, 18 of the largest cities in Israel joined the International Council for Local Environmental Initiatives' (ICLEI) Cities for Climate Protection Program (CCP), and signed the Forum 15 Convention, committing them to reduce Greenhouse Gas (GHG) emissions to 20% below the year 2000 levels. In June 2013, those cities decided to adopt the green building standard as mandatory.

Housing Policies: Free Market Versus Welfare State

Two opposing views prevail in the debate about the role of the state in shaping society and the economy: (1) that which assigns to the state a prominent role in regulating the economy and alleviating social problems, and (2) that which advocates limited state involvement and an increased role for the market. The former argues that the state is well-placed to ensure an acceptable standard of living in society, alleviating social problems and correcting market distortions– assuming that certain essential goods and services, such as housing, cannot be procured by all citizens. Opponents of this "welfare state" approach argue that government involvement in providing goods and services weakens the individual's sense of personal responsibility and distorts market relations.

The leading objective of Zionist regional strategies in pre-state years was the geographical deployment of rural agricultural settlements – often kibbutzim (collective farms) and moshavim (smallholder cooperatives), both of which are located on nationally owned land (Efrat 1984).

The first leaders of the State of Israel were Labor Zionists with a largely socialist ideology, considering housing as a tool of nation-building and developing a just society. This belief, together with a strong conviction in the right and ability of the state to manage all the important aspects in the life of its residents, guided housing policy in the first 20–30 years of Israel's existence (Carmon and Mannheim 1979). With independence came large waves of Jewish immigration from all over the world, and housing policy was meant to achieve immigrant absorption, and the dispersion of the population to the country's geographic periphery, considered essential to national security (Shadar 2004). Availability and the control of the government over crucial development resources turned Israel into an absorption center for a huge mass of Jewish immigrants from all over the world. By 1950 the Jewish population had almost doubled. There was considerable government involvement in the housing market, direct and indirect (Carmon 2001). By the mid-1960s, there had been massive government investments in building public housing, developing public services and institutions, and providing incentives for industrialization and regional infrastructures (Kipnis 1987). The Planning Division of the Israeli Government was established to plan population distribution, and design new settlements and residential neighborhoods (Shadar 2004). The Housing Division, another government body, was responsible for housing construction and shelter provision to new immigrants. The government provided funding for low-cost housing projects, or "shikunim" (Shadar 2004). From the beginnings of public construction, about half the apartments were earmarked for sale to residents, while the rest were designated for public management and rental. The largest public housing company, Amidar, was established as a quasi-governmental entity (and still manages most of Israel's remaining public housing stock, a small proportion of what once existed).

Thus, a high degree of centralization marked Israel's building industry throughout its first decades of statehood. The political upheaval of 1977 ended three decades of Labor hegemony, and privatization measures implemented by the new right-wing coalition were a first stage in reducing government intervention in construction (Rabinovich 2007). Since this turning point, the private sector has become increasingly dominant in the Israeli residential building field (Cohen and Amir 2007), as detailed in Table 2.

Along with the shift from public to private sector domination in Israeli construction, residential ownership strengthened: over 73% of residences are tenant-owned (Rabinovich 2007), as opposed to less than 58% in France (INSEE 2010).

According to Esping-Andersen's classification (1990), France is a welfare regime. In 2013, social expenditure represented 33% of GPD (the highest among OECD countries) – nearly double its value in Israel −15.8% (OECD 2014). France has a long tradition of social and state intervention in the housing market (Table 3): In the 1800s, following the 1832 cholera epidemic, the French government enacted

Table 2 The increasing role of the private sector in building in Israel

1948–64	Government involvement was extensive and public construction -dominant, including residential buildings, physical infrastructures and community services. Seventy-eight percent of new housing units were publicly initiated.
1965	The Planning and Building law was passed, restricting the government's freedom of action. Publicly initiated housing in 1965–79 dropped to 45%.
Early 1970s	Housing policy was sharply criticized by economists. Important changes were introduced, including in the eligibility for subsidized loans.
Since 1977	Following the formation of the first right-wing government, the state reduced its direct involvement in the housing market. Only 20% of new housing units in the 1980s were publicly initiated.
Early 1990s	Government involvement rose again, as the private sector could not meet the demand generated by the sudden influx of hundreds of thousands of immigrants. In 1992, 70% of the new housing units were publicly initiated.
Mid-1990s	The government reduced again its involvement in housing construction. Housing policy underwent privatization, accompanied by the development of a universal support system for needy populations by subsidizing rents.
2000s	The government distanced itself from large social programs. Contracts for public housing projects are tendered, awarded and overseen by the Housing Ministry and carried out by the private sector. The State refrained from funding new social housing construction.

Source: Carmon (2001)

Table 3 The introduction of an enforceable right to housing in France

Subparagraphs 10 and 11 of the preamble to the Constitution, 27 October 1946	"The Nation shall provide the individual and the family with the conditions necessary to their development."
The Quilliot Act, 22 June 1982	"The right to housing is a fundamental right."
The Besson Act, 31 May 1990	"Guaranteeing the right to housing is a duty of solidarity incumbent upon the whole nation."
The Anti-Exclusion Act, 1998	Focuses on housing emphasizing measures to prevent evictions, reform housing allocation, act on empty homes and tackle sub-standard housing.
The SRU (Urban Solidarity and Town Planning Renewal) Act, 2000	Confirms the right to decent housing. In every municipality of more than 3500 residents must include at least 20% of social housing.

Source: Loison (2007)

the law on sanitation of substandard housing, and during the industrial revolution, the private sector established company towns, known as "cités ouvrières" (Alvim and Leite Lopes 1990).

Three periods have been identified in French housing policy following World War II: the first was marked by heavy State intervention in housing production, followed by a re-focusing of fiscal attention on the older city centers and their rehabilitation, and finally by a search for efficiency in financial assistance, while maintaining strong national solidarity in housing (Blanc 2010). In the first years after World War II, France experienced a persistent housing crisis. Although the Fourth Republic successfully carried out reconstruction following the war, this did not substantially reduce housing demand –which was spurred by urbanization, population growth, the repatriation of one million French nationals from Algeria, immigration, and the deterioration of buildings (by the turn of the twenty-first century, 35% of the nation's housing stock predated 1948). The government encouraged construction through premia, loans (particularly for low-rent housing), and tax incentives.

Municipal and other public bodies also engaged in a vast program of subsidized public housing known as "habitations à loyer modéré" (HLM's), especially prominent in the 1960s and 70s. In the 1970s the procedure for receiving building permits for private construction was greatly simplified, and since 1982 mayors have been responsible for granting construction permits and devising local housing policies for both the public and private sectors. The government has also sought to encourage home ownership through low-interest loans. From the late 1960s, city planning in France became more organized, through such programs as mixed development zones or "Zones d'Aménagement Concerté" (ZAC), which often link private and public developers. Reforms in 2000 updated long-term development plans (Schéma de Cohérence Territoriale – SCOT) and detailed land-use plans (plans locaux d'urbanisme; PLU). The current emphasis of urban policy is on rehabilitation, particularly of the many peripheral housing estates built in the 1960s and 70s, but also of older central districts (Lefebvre et al. 1991; Blanc 2010).

'Green' Building Development in Israel

In Israel, a standard requiring minimum levels of thermal insulation in buildings (SI 1045, first issued in 1979) was adopted in 1986 as compulsory within the national planning and building regulations. However, the effectiveness of this standard in improving building energy efficiency has been limited by its modest requirements, as well as by its low level of enforcement (Bar Ilan et al. 2010). In 2005, a voluntary standard known as SI 5281: "Buildings with Reduced Environmental Impacts – Green Building" was launched under the sponsorship of the Israel Ministry of Environmental Protection. Applicable to all new or refurbished residential or office buildings, SI 5281 offers a Green Building Label, accredited by the Israel Standards Institute (ISI) and based on the accumulation of credits in a number of categories as described above. Also, in 2005 the ISI published SI 5282: "Energy Rating for Buildings" and this rating scheme was approved as part of the energy requirements of SI 5281 for residential buildings. In 2007, it was adopted for office buildings as well. Like the green building standard, SI 5282 is a voluntary mechanism and therefore has been considered by many as an insufficient policy instrument for achieving meaningful improvement in building energy efficiency at the national level (Bar Ilan et al. 2010). In Israel, the residential sector accounts for 31% of the country's electricity consumption (Central Bureau of Statistics 2005). According to the Ministry of National Infrastructures (MoNI), educational campaigns in the media could produce a 8–10% reduction in energy consumption in households. In addition, upgrading or replacing old, inefficient appliances could save 25–40% of the energy consumed (Yehezkel 2008).

In September 2008, the Israeli government adopted Resolution 4095 to increase energy efficiency and reduce 20% of electricity consumption by the year 2020, relative to the year 2006 (Prime Minister's Office 2008). The decision includes policy measures focusing on public facilities, government offices and local authorities. For example: The yearly expenses on electricity consumption of each ministry will be pre-determined in its annual budget. As an incentive for applying energy efficiency measures, any budgetary surplus due to the implementation of energy conservation strategies, remains in the hands of the particular unit. In addition, each ministry must audit its energy consumption and survey its conservation potential. Accordingly, it should reduce energy consumption by at least 10% by the year 2020. These decisions are also to be implemented in the Parliament and local governments. Local authorities in the periphery receive special financial assistance for this purpose. The Israel Energy Forum, a leading NGO in the field, criticized this resolution due to its narrow focus on the governmental sector. The forum argued that the program fails to sufficiently address the residential, industrial and commercial sectors, that account for 90% of energy consumption (Sverdlov and Dolev 2009). In July 2010, almost 2 years after the governmental resolution was accepted, the MoNI has published its "National Plan for Energy Efficiency- Reduction of Electricity Consumption 2010–2020" (MoNI 2010). The plan is based upon an analysis of the energy conservation potential in all market sectors.

In 2011, the Green Building Standard (SI 5281, with SI 5282 as a basis for the energy requirements) underwent a comprehensive revision and expansion involving numerous stakeholders. While still denoted as SI 5281, it included vastly expanded criteria and covered seven different building types (residential, offices, educational, tourist accommodation, retail, healthcare and sites of public gatherings). The 2005 version numbered 20 pages, whereas the 2011 version, with all the sections together, reached 1000. Two additional sections were planned but not developed at that stage. In the revised version of SI 5281, points are awarded in categories of energy, land, water, materials, health and well-being, waste, transport, construction site management and innovation, with minimum "threshold" criteria in each category, except for innovation and transport. Buildings are certified according to a graded scale of one to five stars based on a weighting of points accumulated in all categories. This process has increased the visibility of the green building within the local construction industry, and implementation of the standard has been further accelerated since Jun 2013, when it was adopted as mandatory by Forum 15 municipalities, that joined the ICLEI initiative for addressing climate change at the urban level (Goulden et al. 2017).

Thus, green building standards, in Israel as in other countries, are increasingly being adopted as mandatory requirements by the authorities. Furthermore, a set of principles for the assessment of sustainable neighborhoods started to be developed by the Israeli Green Building Council (ILGBC) and other actors in 2013 (Goulden et al. 2015). The assessment tool called 'Neighborhoods 360' designed in 2016 by the ILGBC and the Israeli Ministry of Housing, includes three main categories:

'Infrastructures and Construction, Public and Open Spaces, Efficient Use of Resources'. The first pilot projects, launched in 2016, are ongoing and no case study assessment is yet feasible.

A number of municipalities have initiated the development of self-declared 'green' neighborhoods since the late 1990s. The first of these was a 'green' neighborhood in Kfar Saba, in which planners created an "applicational model for a green neighborhood" (Olander 1999) included as an environmental appendix to the submitted plan and part of the project's specifications. The requirements included measures pertaining to noise prevention, air quality, infrastructure, waste disposal and recycling, and climatological design (2014, Head of Environmental Planning – Sharon region Environmental Unit, "personal communication").

Israel's development of environmental policies, such as in the field of green building, has often taken cues from the Europe – where such policy development is generally seen as being more advanced. Despite the umbrella of EU directives, however, actual policy among member nations is quite diverse, and France is in fact one of the latest European countries to implement a sustainable urban development policy (Lacroix and Zaccaï 2010).

'Green' Building in France

According to Boutaud (2009), the European Union's promotion of sustainable urban development can be credited to a few pioneer countries, such as Sweden, Britain and the Netherlands (Fig. 1).

In France, the HQE (High Environmental Quality) label for buildings was established in 1994. The ratification of the Kyoto Protocol by the French authorities (1997) and the EU Energy Performance of Buildings Directive (2002)[1] led the State to plan a progressive regime of Thermal Regulations (RT2000, RT2005, RT2012) to reduce the environmental impact of building operation. Several French building labels have been created to promote Energy Efficiency improvements. Best known are the High Energy Performance (HPE) label, the Very High Energy Performance (THPE) label and the Low Consumption Building label (BBC, energy consumption ≤ 50 kWh/m^2). Moreover, to encourage people to take environmental concerns and energy savings into account, incentives for the renovation of housing and building heating systems were established in 2005. Defined by the Ministry ruling of May 3, 2007, the generic term of Low Consumption Building (Bâtiment Basse Consommation – BBC) "certifies the conformity of new buildings with thermal regulation, compliance with a global energy performance by the building, exceeding the regulation requirements and the minimum inspection conditions".

---

[1] The European Union has supported the improvement of building energy performance with a range of legislative and funding mechanisms and instruments. A key part of this legislation is the Energy Performance of Buildings Directive (EPBD). The EPBD was approved on 16 December 2002 and brought into force on 4 January 2003. Its principal objective is to promote the improvement of the energy performance of buildings within the EU, through cost-effective measures.

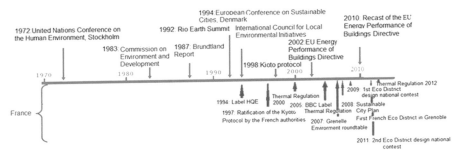

Fig. 1 Timeline of events leading to Sustainable Urban Development and 'Green' Building policies in France (Machline 2014: Own graph)

The BBC-2005 label was included in the 2012 Thermal Regulation, as a supporting document. The Effinergie Association established values for the BBC-Effinergie label, which call for a yearly maximum consumption of 50 kWh of primary energy per square meter of floor area. The consumption taken into consideration is for central heating, domestic hot water, air conditioning, lighting, ventilation and auxiliary heating devices. Airtightness is also required. The BBC-Effinergie label is issued by private certifying firms (Certivea, Cequami, Cerqual and Promotelec). Launched in January 2007, the BBC-Effinergie label goes beyond the requirements set by the RT 2005. Overall, the performance of a certified building is improved by 50% compared to the regulatory base, in terms of energy consumption (50 kWh_{pe}/m^2/year).

By the end of September 2011, a total of 42,000 individual houses, 230,000 collective dwellings and 543 tertiary operations (representing an overall built area of 4 million m^2) had been awarded the BBC-Effinergie label or were in the process of being certified (Boughriet 2012).

At the neighborhood scale, in 2008 the Ministry of Ecology, Sustainable Development and Energy launched a national program of Eco-Districts (Éco-quartier) as part of the ministerial plan "Sustainable City" (Plan Ville Durable). It aims at improving the quality of life, preserving resources and landscapes, and adapting dwellings to residents' new expectations and lifestyles. The Program enhances the need for Eco-quartiers to contribute to Sustainability: Environmental, Social and Economic.

The framework is organized in Four Categories and has been updated every year: (1) Environment and Lifestyles, (2) Territorial Development, (3) Resource Preservation & Climate Change Adaptation, and (4) Approach and Procedures.

The main implementation tools of the program included:

- A Call for Projects in 2009 and one in 2011, offering awards for the best projects,
- A Set of Guidelines organized into four categories to cover the design, construction and management of EcoDistricts and to achieve:

  – Sustainable water management;
  – Optimal waste management;

- Urban Biodiversity;
- Alternative modes of transportation (trams, bicycles, etc.);
- Density;
- Use of eco-friendly materials; and
- Social Diversity and Mixed-use.

- The label "ÉcoQuartier" created in December 2012.
- A national "*Club EcoQuartiers*" with a membership of over 500 local governments, exchanging best practices, training, visits and conferences.

The French government started a national program by creating a network of multiple actors: national institutions (specializing in project funding, affordable housing, environmental protection, and transportation), local governments, construction and utility companies, and citizen associations. The government offered no funding but provided recognition as well as access to shared knowledge and training, including how to obtain public and private financing. Competing projects multiplied from 160 in 2009 to 394 in 2011, for 28 and 24 awards respectively (De Chastenet et al. 2016).

'Green' Value

Improved environmental performance is expected to enhance a building's attractiveness, boosting its monetary value. A growing number of empirical studies demonstrate that the marketing of green buildings tends to raise purchase and rental prices, which causes them to be unaffordable to the local population – resulting in a gentrifying effect.

As pointed by Eichholtz et al. (2010), sustainability pertains to methods of production, as well as qualities of the consumable product and attributes of capital investment. It thus "reflects popular concern for environmental preservation but may also reflect taste changes among consumers and investors". The popularity of the idea of sustainable development led the actors to consider the potential value generated by the environmental performance of buildings, labeled "green value" in the literature. Sustainable performance of buildings is expected to improve their attractiveness, raising their value. The main issue is estimating the value premium generated by these attributes. As detailed in following table, a growing number of empirical studies demonstrate that green buildings allow for rental/sale premia, as well as higher occupancy rates and thus higher asset values (Wiley et al. 2010) (Table 4).

The results suggest that green buildings have a "gentrification effect". Consequently, improving the performance of buildings should raise values for investors or landlords by more than the extra costs of going green (Bartlett and Howard 2000). Many studies insist on the difficult estimation of the actual value created by sustainable buildings. The value premium is often attributed to attractiveness for occupiers due to decreased operating expenses with energy efficiency, or productivity gains and improvement of employee well-being (Kats 2003).

Table 4 Summary of 'green' value studies

Year/Country	Study name	Criteria examined	Results
2009, Brounen and Kok, the Netherlands	"Energy Performance Certification in the Housing Market Implementation and Valuation in the European Union"	Value of dwellings with energy label (A, B and C)	Market value 2.8% greater; Energy criterion has more of an impact than the multi criterion environmental certification
2010, City of Darmstadt, Germany		Energy criteria	+ € 0.38 /m² for consumption. <250 kWh/m²/year + € 0.50 /m² for consumption Ep. <175 kWh/m²/year
2009, Griffin et al., USA (Portland)		Green buildings – Energy Star or LEED certified	Sale price: between 3% and 9.6% higher Selling time / Selling time on the market: −18 days
2011, ADEME, France	Microeconomic study of ADEME [Environment & Energy Agency], Green value for accommodation (theoretical case studies)	BBC new and BBC renovation energy criterion	Renovation: Green value from 5% to 22% of the market value; New multi residential: around 5.5% of construction cost (around € 13,500 per apartment) New detached houses: around 6% of construction cost, but varies depending on the type of energy (wood pellets, gas, electricity)
2008–2010 Salvi et al., Switzerland	Studies by the BCZ (Zurich cantonal bank): Impact of the "Minergie" label on the market value and rental value in Switzerland	Value of dwellings with Minergie label	Individual houses: Market value: +7% Multi-family dwellings: Market value: +3.5%; rental value: + 6%
2010, Marco Salvi, Juerg Syz, Switzerland	What Drives Green Housing Construction? Evidence from Switzerland	Motivations leading to the construction of green dwellings.	Cultural affiliation and income levels of municipalities involved in green construction.
2010, Ben J. Kaufman, U.S.A.	"Green homes outselling the rest of the market"	Energy Star or LEED for Homes certifications or checking by a third party.	In Seattle: Sales of certified new houses: 9.1% more and 4 times faster than non- certified new houses
2011, Brounen and Kok, the Netherlands	"Residential Energy Use and Conservation: Economics, Demographics, and Standards"	Determination of the impact of occupant's behavior on energy consumption	Differences in energy consumption are due to (a) building technical characteristics and (b) occupant profiles and behavior.

(continued)

Table 4 (continued)

Year/Country	Study name	Criteria examined	Results
June 2011, Earth Advantage Institute, Portland (U.S.A.)	"Certified Homes Outperform Non-certified Homes for Fourth Year"	Energy Star or LEED for Homes certifications, Earth Advantage New Homes.	Certified houses: average sale price 30% higher than non-certified ones Certified new houses: price 8% higher than non-certified new houses (sources Earth Advantage Institute) Certified new houses: sale 18% greater than non-certified new houses (sources RMLS Portland)

In attempting to craft green programs for affordable housing, important questions for urban policy-makers include whether green strategies can be effectively integrated into affordable housing developments, what the magnitude of the green construction cost premium is, and whether developers could afford it.

While in Israel there are no explicitly "green" affordable housing projects, in the French Eco-quartiers social diversity is a "programme component" (Souami 2009). Decision makers manipulate methods of funding, the status of the accommodation and the size of apartments, to generate social diversity – and some apartments are rented as public social housing ("HLM"). Moreover, the share of affordable (HLM) housing in "green" developments is often higher than the 20% required by the "Urban Solidarity and Renewal Act" (2000). Thus, the State provides financial incentives to make 'green' building more accessible.

Sustainable Affordable Housing: A Contradiction in Terms?

As pointed by Eichholtz et al. (2010), sustainability concerns methods of production as well as qualities of the consumable product and attributes of capital investment. It thus "reflects popular concern for environmental preservation but may also reflect taste changes among consumers and investors". The popularity of the idea of sustainable development allowed actors to consider the potential value generated by the environmental performance of buildings, labeled "green value" in the literature. Sustainable performances are expected to improve building attractiveness (Bartlett and Howard 2000). The main issue is estimating the value premium of these attributes. As detailed in the previous section, empirical studies demonstrate that green buildings allow for rental/sale premia, as well as higher occupancy rates and thus higher asset values (Wiley et al. 2010). The results suggest that green buildings have a "gentrification effect. Many studies insist on the difficult estimation of the actual value created by sustainable buildings. The value premium is often attributed to attractiveness for occupiers due to decreased operating expenses with energy efficiency, or improvement of employee well-being and productivity (Kats 2003).

Historically, affordable housing development has been characterized by an emphasis on low upfront capital and construction costs (Bradshaw et al. 2005). Achieving affordability by minimizing capital investment, however, has often proved a short-term solution, sacrificing long-term building functionality. When cheaper, lower-quality systems fail over time, building operating costs increase sharply, hurting both building owners and residents (Bradshaw et al. 2005). There are several ways to make green building more affordable: by modifying demand, by reducing the real resource costs of producing housing or by liberalizing the regulatory system to ensure that more building land is made available. Policies to help ensure affordable housing to particular groups, disadvantaged within a market framework, include income supplements, targeted price reductions, or a regulatory mechanism segmenting the housing market. This last policy would aim at ensuring that the demand from higher-income households does not raise prices for poorer groups and enough land is allocated to the affordable housing segment, to ensure lower prices for those with lower incomes. Providing housing at below-market prices cannot be done without additional subsidy, which could reflect the social value of helping poorer groups, or provide a mechanism to compensate for resource mis-distribution (Barker 2004).

In attempting to craft green programs for affordable housing, important questions for urban policy-makers include whether green strategies can be effectively integrated into affordable housing developments, what the magnitude of the green construction cost premium is, and whether developers could afford it.

While in Israel there are no explicitly "green" affordable housing projects, in the French Eco-quartiers social diversity is a "programme component" (Souami 2009). Decision makers manipulate funding methods, accommodation status and apartment sizes s, to generate social diversity – and some apartments are rented as public social housing ("HLM"). Thus, the State provides financial incentives to make 'green' building more accessible.

Towards Eco-Gentrification

Gentrification is a process of urban transformation, whereby the population of a local community is displaced by a higher income population. According to Clark (2005): "Gentrification is a process involving a change in the population of land-users such that new users are of higher socio-economic status than the previous users, together with an associated change in the built environment through reinvestment in fixed capital."

Gentrification occurs in various ways in different neighborhoods of different cities, comprising diverse trajectories of neighborhood change and implying a variety of protagonists (Lees 2000). For gentrification to occur, it must be initiated by "some form of collective social action at the neighborhood level" (Smith 1996). Key actors include residents, local governments, financial bodies, and real estate professionals.

Gentrification has become a global phenomenon as well as an increasingly important strategy. Local governments may favor it to improve their tax base, as well as for other perceived benefits of moving poor people out.

Gentrifying neighborhoods are typically characterized by upward pressure on housing prices. There may be different effects on renters and homeowners, and varied consequences for different homeowners. The increase in property values may settle at a new high or reflect "unsustainable speculative property price increases" (Atkinson and Bridge 2005). Often there is a loss of affordable housing, particularly in the rental market, which can be exacerbated by zoning changes that eliminate single-room occupancies or other low-cost alternatives. Thus, the rise in property values can be fortunate for families who owned homes, but devastating to renters, although homeowners may struggle if their incomes cannot keep pace with rising property tax bills and may find themselves required to sell their home (Slater 2004). For example, in the Brooklyn Heights neighborhood, a house purchased for $28,000 in 1962 (or 140,000 in $ of 1995 was sold for $640,000 in 1995 – nearly 5 times more (Lees 2003).

In Israel, at the edges of large cities, older neighborhoods are changing, as well-off professionals move in and raise housing prices. In Tel Aviv, for the past several years, residents of the older southern neighborhoods have seen an influx of young people from central and north Tel Aviv earning above-average salaries (Marom 2014).

This process is occurring in most south Tel Aviv neighborhoods, such as Shapira, Neveh Sha'anan and, of course, Jaffa's seashore, a trendy new location for Tel Aviv's middle class. In less than a decade, housing prices have risen by over 100% in these neighborhoods, and some have seen major construction and urban renewal projects. The same goes for places such as Neveh Sharett (northeast Tel Aviv), Morasha (Ramat Hasharon), Neveh Yisrael and Neveh Amal (both in Herzliya), and many others around the country – nearly all in high demand areas. In Neve Sharett, for example, it was possible to buy a 3-room apartment in 2008 for 400,000 shekels [about $100,000], while in 2014 it costs 1.5–1.6 million shekels. Along with the higher standard of living and improved infrastructure that accompany this process, old residents are priced out of the neighborhood.

In neighborhoods where gentrification has been extensive, "original residents" are hard to find. This is the case in Neveh Tzedek and in nearby Florentin in south Tel Aviv (Alexander 2012), as well as in Jerusalem neighborhoods such as Ein Karem, Nachlaot and the German Colony. Before they were transformed into prestigious neighborhoods, all included lower-middle-class residents (Gonen 2015). According to the head of the Tel Aviv municipality planning unit (2017, "personal communication"):

> One of the biggest problems in urban renewal is the ratio between the future number of units to the present one… For the project to be economically [feasible], this must be maximized. This means, in almost all cases, that tower blocks with high maintenance costs replace older buildings… In many cases, decision makers see the entry of a new and 'high-quality' population into a troubled neighborhood as a goal... because as far as they're concerned, that's how they're strengthening the neighborhood… What usually happens is that the neighborhood changes completely and the new population takes over... The cities

encourage a mix of housing in the name of integration and preventing concentrations of poverty, and in the name of this goal, a strong population is 'imported' into a weak neighborhood. But the opposite doesn't happen. No municipality thinks they must create a mix in richer neighborhoods, which will allow weaker groups to live there. It's always one-way.

According to the Head of the Strategic Planning Department in Tel Aviv's municipality (2016, "personal communication"): "No magic solution exists to completely prevent gentrification, it happens in every large city worldwide… there are only ways to try to regulate the process. Today, we are much more aware of the problem than in the past and are trying not to plan tall towers in weaker areas. It's a complex and challenging situation … All parts of the city are entitled to development, infrastructure and a supply of new and larger housing, just as the richer neighborhoods are too – and the minute you build such projects, a risk of gentrification exists'… In Tel Aviv, they are now setting up a legal basis for funds to help with the maintenance costs in apartment blocks for the less well-off in areas that underwent urban renewal, allowing them to stay in their old neighborhoods."

Eco-gentrification or environmental gentrification is recent. It is the result of urban policies inextricably linked to economic development. Following Checker (2011), we understand environmental gentrification as:

> …the convergence of urban redevelopment, ecologically-minded initiatives and environmental justice activism in an era of advanced capitalism. Operating under the seemingly a-political rubric of sustainability, environmental gentrification builds on the material and discursive successes of the urban environmental justice movement and appropriates them to serve high-end redevelopment that displaces low income residents.

Cleanup makes a neighborhood more attractive and may drive up real estate prices and rents, displacing local working-class populations who have suffered the consequences of economic trends like deindustrialization, while richer homeowners capture the gains in their property assets (Banzhaf and McCormick 2007). While sustainability and green urbanism have become central in urban policies, too little analysis has focused on who gets to decide what 'green' looks like. Many visions of the green city seem to have room only for park space, waterfront cafes, and luxury LEED-certified buildings, prompting concern that the "sustainable" city has no room for industrial uses or for the working class (Krueger and Agyeman 2005).

The sustainable city is seen as desirable, but not socially neutral if the densification and regeneration of central areas only benefits a well-off minority. Emelianoff (2007), for example, found that eco-neighborhood projects in Europe usually focus on the middle to upper classes, given the costs of renting or buying project dwellings. Theys (2002) also argues that the 'zero-default city' (with a very high quality of life) would be financially inaccessible to most people, thus deepening social inequalities.

Decamps (2011) has analyzed the impact of sustainable principles on corporate property decisions and on the attractiveness of business districts in France. The results confirm that the impact is positive and strong. Modern amenities coupled with high-valued metropolitan functions may improve territorial attractiveness, encouraging the settlement of high-income residents and thus gentrification.

In Israel, 70% can't afford the average price of an old apartment, while 80% can't afford a new one (Milken Institute 2015). The social protest of 2011 expressed the economic distress of individuals and families in the mainstream of Israeli society – in particular, the distress of young and working families, with appropriate education and employment qualifications, who are overwhelmed by the costs of living, of housing and of the proper care and education of their small children. It focused explicitly on housing unaffordability. Housing has become the decisive component of the relationship between state and citizen in Israel, capable of drawing hundreds of thousands into the streets (Trajtenberg Report 2012). "The people demand, social justice!" The slogan was voiced by hundreds of thousands in the squares of Israel's cities. Recent statistical data show that as of 2011, an Israeli homebuyer needed the equivalent of 121 months' worth of average wages to purchase a four-room apartment – while in the U.S only 60 were required and in France 90 (OECD 2013). Considering that less than 0.04 public social housing units exist in Israel per 1000 habitants (Paz-Frankel 2012; see data in annex) – a small fraction of what has been allocated under the "socially minded" regime in France – it is timely to examine the social and economic implications of an initiative that is gaining momentum in the name of environmental "sustainability". With this in mind, our study aims to identify to what extent green building is fostering eco-gentrification in Israel, and what lessons can be learned from the parallel, but distinctly different, history of green building in France.

Sustainable Urban Development and Social Justice

Following the WCED (1987), "sustainable development" has become a widely accepted conceptual framework for urban policy and development, providing context to a considerable literature on planning, architecture and urban design (Williams et al. 2000). Debates about sustainability are no longer limited to environmental aspects, but incorporate economic and social dimensions (Fig. 2).

While the social dimension of sustainability raises questions of social justice within the goals of sustainable development, there is no agreement on answers (Hopwood et al. 2005) – despite a recent European policy focus on 'sustainable communities' and social cohesion. The 'Bristol Accord' (2005) spells out a common European approach to 'sustainable communities' signed up to by EU member states, building on previous EU initiatives, including the Aalborg Charter and Agenda 21.

Fig. 2 The three spheres
of sustainability

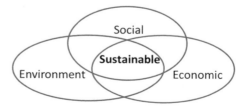

The concept of social equity within definitions of sustainable development (Hopwood et al. 2005; Chiu 2002) has focused on meeting the needs of present as well as future generations (WCED 1987; Holden and Linnerud 2007), to redress inequities of outcome (Haughton 1999). An equitable society is one with no 'exclusionary' or discriminatory practices hindering individuals from participating in it economically, socially and politically (Pierson 2002). For measuring social equity, accessibility is a frequent criterion (Burton 2000).

Social equity is directly linked to the built environment, either through the provision of services, or by the means of accessing them (e.g. public transportation). Additional criteria include access to decent housing, as measured by its physical condition and by the services provided by the relevant housing association/local authority. Furthermore, problems of housing affordability (and tenure) may prohibit residents from living in or moving out of neighborhoods and areas. Prompted by an entrepreneurial mode of governance, sustainability has increasingly become a buzzword for urban environmental and political governance (Keil 2005).

The surge of environmental awareness and regulation, has created 'green' markets, especially in real estate development, leading to the "eco-gentrification" of some neighborhoods (Krueger and Savage 2007). Urban renewal, and more recently projects marketed as "sustainable" neighborhoods, can in fact foster social polarization. One goal of such projects is improving the quality of life in a particular location, which means making some spaces more attractive than the rest. If located in an already attractive (i.e. expensive) area, such improvement may paradoxically generate discrimination in access to "sustainable" housing. For example, a major objective of sustainable urban development is discouraging suburban sprawl, through densification of the city center. This strategy, grounded in ecological motivations, can inadvertently present challenges in the attainment of social objectives, as it tends to increase the attractiveness of these areas, while preventing deprived people from accessing them (Checker 2011).

In France, the State and local authorities encourage projects promoted as sustainable neighborhoods and including 'green' certified buildings. This approach is consistent with the French culture of urban planning, as public authorities are the central actors in urban projects (Souami 2009). In contrast, well-known sustainable neighborhoods in Europe (e.g. Malmo in Sweden, BedZed in England, and Vauban in Germany) are the result of private initiatives by dwellers, property developers or associations. However, a process of innovation has emerged with the "Grenelle"1 (in 2008) and 2 (in 2010) environmental initiatives. In fact, debates organized by the Ministry of Environment with associations, researchers, and professionals have aimed to establish a coherent strategy for sustainable development. Even in France, however, private developers and builders integrate environmental concerns and requirements into their projects. This is intrinsically justified by the social impact and in the reputation of the company.

One of its prominent elements is the "Sustainable City Plan", which contains calls for neighborhood development projects known as "Eco-quartiers" (eco-districts). In promoting eco-quartiers, the ministry has emphasized that the French approach does not neglect the social dimension in sustainable neighborhoods, which is conspicuously absent in other countries. In the French conception, these neighborhoods should be mixed, or socially diverse. As explicitly stated, the issue is to avoid the creation of "ecological enclaves for the upper middle class" (Charlot-Valdieu and Outrequin 2009).

In the following chapters we will present how green building in these two countries, Israel and France, is used – purposefully or inadvertently –to promote social and economic goals. We examine how green building, under the banner of sustainability, may increase inequality in Israel – through "eco-gentrification". In addition, we use data collected in France to gauge the actual effectiveness of its declared policies for promoting social diversity in green housing. We suggest that an in-depth analysis of the social aspects of green building in both Israel and France, can help identify opportunities and obstacles to its successful implementation, while gauging its broader and longer-term effects on society.

In the coming chapters we present case studies in which we look whether social aspects are integral to green building in Israel and in France, by: (1) comparing the policy context and 'green' building instruments used in France and in Israel, (2) investigating whether green building fosters "gentrification" intentionally or unintentionally.

The Socio-Economic Impacts of 'Green' Building in Israel: Green Building as an Urban Branding Tool

The 'Green' Premium in Israel

'Green' building may involve additional construction costs, such as those required for better wall insulation or upgraded window glazing. However, the 'green' premium, or the increase in a property's selling price, may not only include these extra costs but also the extra profit to the developer from 'green' construction (Massimo 2012). Fuerst and McAllister (2011) found that rental prices of regular commercial buildings are lower by 4.1% on average compared to those complying with LEED and Star Energy standards.

In Israel, economic research about 'green' building has only dealt with its costs to builders. Kot and Katz (2013) studied two buildings built according to SI 5281 in Nes Ziona and Netanya, aiming to shed light on the added costs of 'green' components. Their findings indicate an addition of 2.1–4.1%. Gabay et al. (2014) found added costs in the construction of a 'green' office building, between 4.3% and 11.6%, with over 75% of this spent on energy saving improvements and only 4% on certification costs. Beyond these extra costs to the builder, we seek to estimate the added price for the buyer of 'green' residential buildings.

The effect of "green" buildings on real estate prices in Israel was assessed by calculating the "green" premium (i.e. the added financial value of a certified green real estate asset, when compared to a similar non-green real estate asset). We compared the market prices of new apartments in certified green buildings with the prices of similar non-certified apartments (located at a maximum distance of 200 meters) and sold in the same year. The "green" premium was calculated per square meter of dwelling unit in each municipality and for each year, using a list of 250 residential buildings that have received the SI 5281 certification from the Standards

Institute of Israel (SII) (updated 17/07/2016).[1] We included in the sample only multi-family apartment buildings, and ultimately the comparison includes 91 sets of green vs. conventional residential buildings (see list of cities in the results part).

The details of individual buildings were entered in the Israel Tax Authority website, to retrieve the actual sale prices of apartments. These data are based on the declared transaction price and do not include the subsequent cost of apartment upgrade. To identify the corresponding non-certified buildings for comparison with green buildings in the sample, we used the "GovMap" GIS software (http://www.govmap.gov.il) and compared apartments sold the same year and built during the same period (i.e. after 2008 – see map and Table in Annex).

The sample size is currently limited in terms of range, number of projects within a city and number of cities, but since the volume of 'green' residential building is expected to grow, future research using a larger database will become feasible, allowing for comparison of the green premia for buildings at higher standard levels, (planned in the Forum 15 cities over the coming years), and for the examination of 'green' premium changes over an extended period of time.

The 'Green' Premium

According to the calculation method described above, the average 'green' premium was estimated as a percent increase for a number of urban localities across Israel, with the results summarized in Fig. 1 and Tables 1, 2, and 3.

This sample, the 'green' premium ranges between 3% and 14% depending on the city (see calculation details in the Annex). It is highest in the Northern periphery (13%) and the average is 7.3%. In the U.S., we can see slightly higher average values: for example, Khan and Kok (2012) found a green premium of 9% in California, and according to Kaufman (2010) the premium in Seattle is 9.1%. Thus, despite the wide range of values, the average "green" premium in Israel is just slightly below those observed in the U.S.

The 'green' premium is smaller in the center of the country, where most construction takes place. This is because around the Tel Aviv metropolitan area (including the central and Jerusalem districts), housing prices are high regardless of 'green' certification, and in fact the 'green' premium embodies a minimal difference in actual construction costs – given the negligible difference in requirements (Goulden et al. 2015) between SI 5281 at a one star level (Note 2), and the mandatory standard SI 1045, in the climate of the mild coastal region.

---

[1] We did not include Yavne in the calculation of the average green premium (as no conventional buildings have built near the 'Green Yavne' development, which thus lacks a reference sub-sample), and we similarly refrained from including in our sample the seven certified buildings of Ashkelon, as there are no new conventional apartment buildings nearby. We also removed from our sample the one residential building (in Hod Ha'sharon) that received the '2 star' level of green certification.

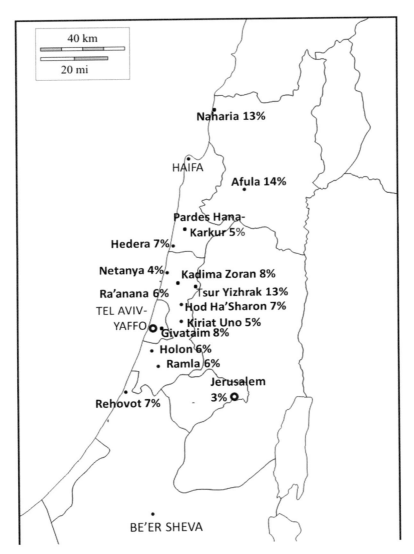

Fig. 1 The 'green' premium in Israel. (Source: Own data (2016), and Map: Machline et al. (2018))

Based on these results, we can estimate the added profitability of green building for developers, by comparing the 'green' premium that the homebuyer will pay with the extra building costs for the developer. According to Kot and Katz (2013), the additional costs of 'green' building construction for new apartments range between 2.1% and 4.1%. However according to the Israeli Builders Association (2015), actual building construction typically represents only 35% of the total project cost (which also includes the costs of land, infrastructure development, design fees, taxes, etc.

Table 1 Number of housing units by city, in certified and non-certified buildings

City	IS 5281 certified buildings Number of units	Non-certified buildings Number of units
Be'er Yaakov	174	134
Givataim	39	31
Hod Ha'sharon	57	28
Hedera	19	14
Holon	41	41
Jerusalem	58	38
Pardes Hana-Karkur	29	78
Naharia	15	18
Netanya	92	93
Zur Yizhak	95	179
Kadima Zoran	7	61
Kiriat Ono	140	42
Rehovot	34	33
Ramla	100	129
Ra'anana	14	16
Afula	41	41
Total	955	976

Table 2 Average price of housing units (4.5 rooms) in certified and non-certified buildings sold in 2013

	IS 5281 homes	Non-certified homes	Difference
Average apartment price (NIS)	1,855,000	1,600,000	255,000
Average apartment size (m^2)	117	120	−3
Average price per unit floor area (NIS/m^2)	15,793	13,642	2151

which are not likely to vary significantly due to the building's 'green' design, certification and construction). This means that the average 'green' premium of some 7% that a homebuyer will pay for a new apartment is considerably higher than the percentage added cost to the developer – which is in fact marginal in the scope of the overall project cost (around 1%). If, as stated by the Israeli Builders Association (2015), the average profit for a conventional apartment sale is 12.7% of the housing unit price, the profit for a 'green' apartment is likely to be in the range of 15–25%.

The Economic Impact of 'Green' Building in Israel: Prices and Profits

Does 'green' building raise housing prices? Do homebuyers profit from purchasing a 'green' apartment? In short, does it provide 'green value'?

Table 3 The 'green'
premium in Israel

City	'Green' premium (%)
Be'er Yaakov	7
Givataim	8
Hod Ha'sharon	7
Hedera	7
Holon	6
Jerusalem	3
Pardes Hana-Karkur	5
Naharia	13
Netanya	4
Zur Yizhak	13
Kadima Zoran	8
Kiriat Ono	5
Rehovot	7
Ramla	4
Ra'anana	6
Afula	14
Average	7.3

Source: Own data (2016)

If we define 'green value' as the added financial value of a real estate asset due its greening, when compared to a similar non-green asset, we find that such added value may be attributed to both environmental benefits, and to less tangible benefits such as added prestige.

In this study we have quantified 'green' value in Israel, by comparing the prices of apartments in SI 5281 certified 'green' buildings (up to 2016), to those of similar apartments in conventional buildings (located in the same municipality, and within 200 meters, of their 'green' counterparts). We found that 'green' certified construction in Israel raises apartment sale prices by 3–14% (about 7% on average), with higher values in the northern periphery than in the country's center (i.e. the Tel Aviv metropolitan area). These 'green premia' in terms of sale prices may be compared with the added construction costs of these 'green' buildings, which by one estimate (Kot and Katz 2013) are around 3%. Note again that Israel's housing market has seen excess demand in recent years (Dovman et al. 2012), and thus every buyer of a 'green' apartment did not necessarily select it over an equally available conventional apartment for its 'greenness' – but rather as part of a decision-making process involving numerous variables (including location, size, etc. as well as price).

We may ask who profits from 'green' building, starting with construction companies and developers. According to the Israeli Builders Association[2] (2015), actual building construction represents on average 35% of a residential project's total cost. Furthermore, 'green' building added costs represent about 1% of the total project cost. This is much less than the 'green' premium that a homebuyer in Israel is likely to pay, which according to our analysis averages 7.3%. Even allowing for uncertainty and variation between parts of the country, the evidence suggests that developers are making a significant added profit on green apartments, which represents the bulk of the 'green' premium paid by homebuyers.

The Israeli Ministry of Environmental Protection claims that 'Green building may save about 15% of electricity consumption and 10% of the water demand in a household'.[3] An Israeli household consumes about 7700 kWh on average per year and the electricity price is 0.6 NIS/kwh (Israeli Electricity Company 2011). Thus, thanks to 'green' building an Israeli household may consume about 1150 kWh less, representing a savings of about 700 NIS ($200), per year. An Israeli household also consumes 181 cubic meters of water on average per year, and the average water price is 12.82 NIS/m[3, 4] Thus, 'green' building can reduce consumption by some 18 cubic meters, representing 230 NIS ($66) per year. In sum, an Israeli household can expect a 'payback' of circa 1000 NIS (about $280) per year through 'green' building.

Meanwhile the average price for a 'green' apartment in our sample is 1,850,000 NIS, and assuming an average 'green' price premium of 7.3%, the added expense incurred by the homebuyer for an apartment in a 'green' certified building is some 135,000 NIS. While this illustrative calculation embodies uncertainties and variations –it shows that the time required for a homebuyer to repay the investment in a 'green' apartment is likely to be over 100 years. Thus, the energy and water savings potential of a 'green' apartment fail to justify its purchase.

However, the purchase of a 'green' apartment may well be justified by its resale value. Studies in the U.S. have examined certified 'green' apartment resale values and shown that in Seattle, for example, "green apartment" prices were higher by about 9% and sold four times faster than non-certified homes (Kaufman 2010). Regarding the Yavneh 'green' neighborhood, new apartments sold for 1,200,000 NIS in 2010 are now sold for about 1,900,000 NIS. Their prices are higher than those of non-'green' apartments nearby by about 10%.

---

[2] Founded in 1949, the Israel Builders Association is the sole representative-organization of businesses in the residential, non-residential and infrastructure construction sector. The Association strives to promote the building and infrastructure sectors in Israel, the interests of contractors and builders, and to solve professional issues. The Israel Builders Association claims over 2000 members.

[3] Based on estimates of the Inter-Ministerial Committee for Reducing Greenhouse Gas Emissions.

[4] Per capita water consumption; the average number of persons per household (Central Bureau of Statistics 2014); water price.

Moreover, if developers were only charging consumers for added construction expenses (about 1% of the total project cost) and considering that the standard developers' profit is 12.7% (of the apartment price-Israeli Builders' Association 2015), the 'green' premium for the consumer would only be about 20,000 NIS. Under such conditions, energy and water savings alone would allow a household to repay its 'green' investment in some 20–25 years – a lengthy period, but a relevant consideration for some homebuyers, and thus encourage 'green' building as an investment in environmental value rather than just in expensive real estate.

Finally, we may consider the benefit of 'green' building for the larger Israeli economy. According to Ariav and Amir (2011), the cost to society of air pollution from fossil fuel combustion is 0.18 NIS per kWh of electricity generated, and the electrical consumption of desalinated water is 4 kWh per cubic meter (Israel Water Authority 2009). So given the average annual consumption of water and electricity, each household preferring to buy an apartment in a green building "saves" the national economy some 200–250 NIS/year in pollution costs, in addition to the savings mentioned above (about 1000 NIS) accruing personally to that household.

In coming years, about 45,000 housing units per year are to be built to meet the needs of the Israeli population (National Economic Council 2014). Thus, according to the above estimates, if all new residential buildings were built according to the 'green' standard, the economy would save about 50 million NIS/year. Even though this represents an extreme scenario, most building construction (as of 2017, about 67%) is taking place in the Forum 15 municipalities – where the 'green' building standard has become mandatory for residential buildings.

According to Cohen et al. (2017), improved 'green' building, meeting the stringent requirements of 'two-star' and above ratings, reduces energy consumption and greenhouse gas emissions, and could be rewarded through State subsidies – and given the potential economic benefits, such a policy would appear cost effective.

The average price for a 'green' apartment in our sample is 1,850,000 NIS. Even at the low end of the 'green' premium range, the additional cost of 3% paid by the homebuyer for 'green' certification is in excess of 50,000 NIS, which is unlikely to be recouped within the period of that buyer's apartment tenure. Thus, it is not profitable to purchase a 'green' apartment for its energy and water saving potential. If, however, developers were only charging consumers for added construction costs (about 1% of the apartment price), the 'green' premium investment could more realistically be recovered.

The benefits of 'green' building may also be estimated for the Israeli economy as a whole. According to Ariav and Amir (2011), the costs to society of air pollution due to power generation are 0.18 NIS per kWh of electricity generated, and the electrical consumption of desalinated water is 4 kWh per cubic meter (Israel Water Authority 2009). So, given the average annual consumption of water and electricity, a household purchasing a green apartment "saves" the national economy some 200–250 NIS/year in pollution costs, in addition to the savings (about 1000 NIS) accruing directly to that household.

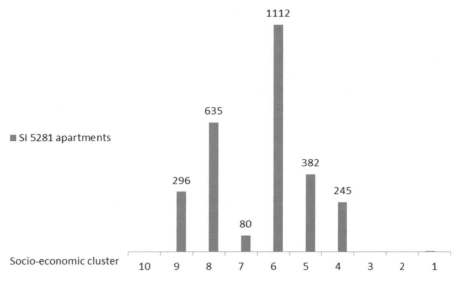

Fig. 2 Number of certified 'green' apartments by socio-economic cluster (2012–2014). (Source: Ministry of Environmental Protection (2016): Personal communication)

In Israel, 'green' apartments are mainly located in high socio-economic cluster municipalities (see Fig. 2). Three quarters of the SI 5281 apartments are located in socio-economic cluster localities of 6 and above, while the populations in the three lowest clusters entirely lack 'green' housing opportunities. Most SI 5281 apartments are located in the center of Israel, where housing prices are highest. Thus, 'green' building is mainly raising the price of real estate in locations where it is already expensive (Fig. 2).

Case Studies

Case studies can be used to test theories or use theory to deepen our understanding (de Vaus 2001), as we have tried to do in the present study. We have looked at a number of 'green' neighborhoods in Israel and France:

1. The Tel Aviv metropolis – (1) Neve Sharett – Green park (SI 5281 certified buildings); (2) Tel Aviv 3700 (A LEED ND planned neighborhood including SI 5281 buildings)
2. Cities in peripheral regions: (1) Yavneh 'green' neighborhoods (SI 5281 buildings) (2) Dimona (A 'green' neighborhood without formal certification).

Case Studies in Israel (Fig. 3)

Fig. 3 Location of the three case study cities in Israel. (Map: Machline (2016))

'Green' Neighborhoods in North Tel Aviv

The neighborhood of Neve Sharett is located in northern Tel Aviv (see Fig. 4).

Originally Neve Sharett was a transit camp, founded in the 1950s to house immigrants, primarily from North Africa. In the 1960s the state built social housing units to replace the transit camps, and public housing tenants have mostly purchased their own homes since the late-1990s (due to the 'Ran Cohen' law (Note 5) of 1998) (Figs. 5 and 6).

Before gentrification, Neve Sharett was a residential neighborhood and included a high-tech center. It stood in stark contrast to the wealthier neighborhoods adjacent to it in north Tel Aviv, and still maintained relatively cheap housing – a disappearing commodity in most of the city. According to the socio-economic data in Figs. 2 and 4, we can see that in Neve Sharett the share of the population holding bachelor's degrees and employed in white collar jobs was very low compared to the rest of northern Tel Aviv, and that the population fell into low socio-economic clusters (between 1 and 5).

In 2003, the Ministry of Housing declared that the neighborhood was eligible for a renewal program. Urban renewal in Israel generates new construction in

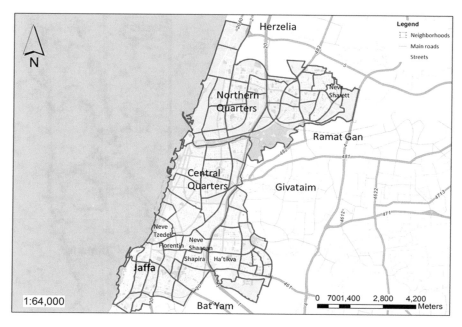

Fig. 4 Division of Tel Aviv into neighborhoods. (Source: Cohen and Margalit (2015))

pre-existing built-up areas, a solution which the government adopted in response to the scarcity of land for new construction and the subsequent housing shortage.

There are two main schemes for urban renewal projects. The first is known as *'pinui-binui'* (evacuation and reconstruction), and the second is an outcome of the national plan to strengthen older structures, known as TAMA 38. Neve Sharett was declared eligible for *pinui-binui.*

Evacuation and reconstruction projects rely on the 2002 Planning and Building Law. Under that law, a specific urban area is declared an area of evacuation for reconstruction – existing buildings may be replaced by new construction, with significantly enhanced building rights (several times greater than present ones), while requiring more green areas. The national *Pinui-Binui* project offers planning grants and tax waivers to developers willing to take the risks, to assist with obtaining owner approval, rehousing during the construction period and construction costs. To encourage *evacuation and reconstruction*, homeowners are offered public grants and tax benefits such as exemption from capital gains tax, purchase tax, VAT on construction services, permit fees, improvement levies, and discounts on municipal taxes in their new apartments.

Typically, the areas declared as sites of *evacuation and reconstruction* are considered under-populated; they often have outdated infrastructures and face significant socio-economic challenges (as is the case in Neve Sharett). When the renewal process is initiated, residents are notified that they live in an area designated for *evacuation and reconstruction*. Homeowners with an understanding of

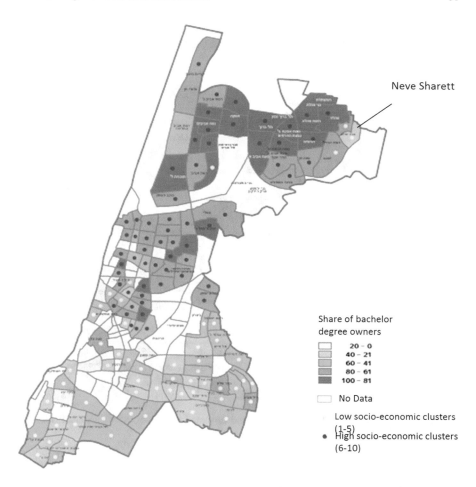

Neve Sharett

Share of bachelor
degree owners

- 20 – 0
- 40 – 21
- 60 – 41
- 80 – 61
- 100 – 81

☐ No Data

Low socio-economic clusters
(1-5)
• High socio-economic clusters
(6-10)

Fig. 5 Map of Tel Aviv census tracts showing the distribution of population in terms of socio-economic indicators (share of population with higher education, and socio-economic clusters)

the issues at hand may rally together and initiate meetings with developers, to either advance the project or prevent it (so as to avoid the eventual inconvenience). In some cases, a developer informed that a specific site has the potential to undergo *evacuation and reconstruction* will approach the residents and attempt to convince them to sign up for the re-construction project. The scheme requires the signed agreement of 80% of the local residents (i.e. apartment owners). In 2006, a reform in the *evacuation and reconstruction* Act was introduced, stipulating that residents refusing to accept the offer of *evacuation and reconstruction* without any clear or reasonable explanation are responsible for damages caused to other residents in the building due to a delay or to the cancellation of the project. (Note 6)

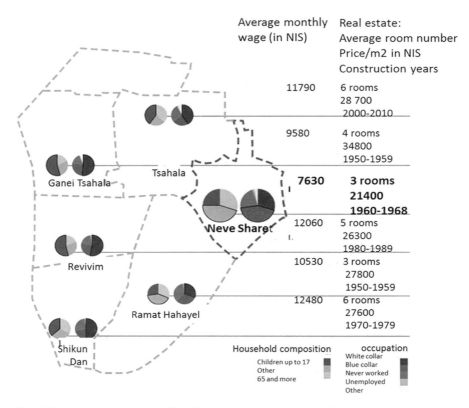

Fig. 6 Socio-economic data for Neve Sharett and surrounding neighborhoods. (Source: Tel Aviv Municipality (2013), Personal communication)

Neve Sharett has been divided into a several *evacuation and reconstruction* projects. We focus here on one of them, the 'Green Park' project, in which the buildings were certified by a private institute rather than by the SII itself (SI 5281, at the base level of one star).

The Green Park project was initiated in 2006, when the local residents hired professional consultants (an architect and a lawyer) to defend their rights and establish a master plan for their neighborhood under an *evacuation and reconstruction* framework. The neighborhood's 154 households (as of 2013) included 62 private rental tenants, 16 public social housing tenants and 76 owner residents (Fig. 7).

The consultants declared that they work with a "social model" through the following steps:

1. The apartment owners contact the consulting firm.
2. The firm collects information on the objectives of the owners (including the social housing company Amidar, which owns 16 of the apartments).
3. The consultants establish a master plan for the municipality.

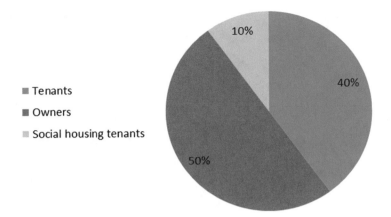

Fig. 7 Residents' categories before evacuation in 2013. (Source: Original survey data (2017))

4. When the master plan is approved, the consulting firm publishes a call for tender to find a developer (stipulating that construction will be according the SI 5281 standard).
5. The developer must get the apartment owners' agreement to start the *evacuation and reconstruction* project (Consulting firm 2017).

The Green Park master plan was accepted by the Tel Aviv municipality in 2013, and three developers were selected to implement it. According to the consulting firm (2017, "personal communication"), the residents wished to include the SI 5281 certification, as they wanted to live in a 'green' neighborhood – although the residents' association representative was unaware that the new apartments were built according to that standard, and simply 'wanted the best building quality standards' (2017, "personal communication"). Thus, one of the requirements published in the call for tenders was to build according to SI 5281 – even though when the plan was accepted, the standard had not yet become mandatory in Tel Aviv and in the other Forum 15 cities. According to the consulting firm (2017, "personal communication"), potential developers were not disturbed by the clause, which is similar to the mandatory SI 1045 standard. According to the arrangements described above, the scheme required the signed agreement of the current apartment owners. The residents lived for decades in ten linear buildings *(binyanei rakevet* (Note 7)), with a total of 154 50-year old apartments. Under the *evacuation and reconstruction* scheme, the ten buildings were demolished in 2014 and six towers, with between 10 and 26 stories and containing a total of 447 apartments, replaced them. The floor area of the new apartments is 120 square meters on average, but 20% are 'small' apartments of 90 sq.m. The occupants (owners and Amidar social housing tenants) were moved to similar rental apartments funded by the developers in close vicinity of their homes. The project developers managed to get planning permission to build three new apartments for each existing one (Fig. 8).

Fig. 8 View of the site following completion of the project, known as "Green Park" due to its 'green' certification (SI 5281). (Photo: Machline (2016))

Economic Impact

1. For the individuals

Each owner received a brand new, 4.5 room apartment (120 sq.m. on average). All apartments include modern kitchens and bathrooms, a balcony and two covered parking spaces. Each new building showcases a luxurious lobby, a state-of-the-art elevator, and manicured gardens. The market price of these new apartments is at least 2.3 million shekels (whereas existing units, prior to the renewal, were valued on average at 1.1 million shekels).

2. For the developers and the consultants

Following the redevelopment process, the developers' investment yielded 293 new apartments to sell on the free market at full price, while another 154 units were allocated to former apartment owners.

The lawyer for the consulting firm that drew up the master plan received a commission equal to 1.5% of total apartment sale revenues, higher than typical commissions which are currently between 0.75% and 1% ('Calcalist' 2016). The architect who designed the project received 10,000 NIS per apartment and another 40 NIS/sq.m. for underground space, 200 NIS/sq.m. for commercial buildings and 400,000 NIS when the master plan was accepted. Hence the architect received total compensation of about 17,000 NIS per apartment, as compared with about 4500–6000 NIS usually paid for such work ('Calcalist' 2016). The consulting firm in charge of collecting the agreement forms of residents (owners) received a commission equal to 3.5% of sales revenue. Considering that apartments were sold for 2.5 million NIS on average, this commission amounts to about 25.6 million NIS ('Calcalist' 2016). By July 2017, all of the apartments marketed as 'green' had been sold (for prices between 2.2 and 4.5 million NIS).

Social Impact

1. *Population displacement and social crisis* (Fig. 9)

All (non-social housing) tenants who were renting an apartment in the neighborhood by 2013 (40% of the residents) had to leave by the end of their contract, and none purchased an apartment in the new project or could afford to rent in the renewed neighborhood (2017, residents' union/social worker, "personal communication"). According to the municipality social worker (2017, "personal communication"), those who were renting an apartment in Neve Sharett will be displaced, and "due to gentrification and rising real estate prices, will find no apartment in the neighborhood, nor in the rest of Tel Aviv, and will thus have to leave the city".

Among the 60% of the residents entitled to an apartment (the residents-owners and public social housing residents) in the new project, the following survey (March 2017) aimed to find out whether they plan to live in 'Green Park' (Fig. 10):

According to the survey, only 25% of the residents that will receive a new apartment will actually live in the neighborhood following the *evacuation and reconstruction* project. Thus, the 'Green Park' project of Neve Sharett is clearly leading to population displacement – symptomatic of urban gentrification. At the same time, the process of *evacuation and reconstruction* offered existing apartment owners "not living in Neve Sharett, because of its bad shape and reputation, the opportunity to return to a newly refurbished neighborhood" (Residents' Association Representative 2017: Personal Communication).

2008: 400 000 NIS   2016: 2 500 000 NIS

Fig. 9 Green Park neighborhood: 'social model' or 'greentrification'? (Source: Tel Aviv municipality/Developer website)

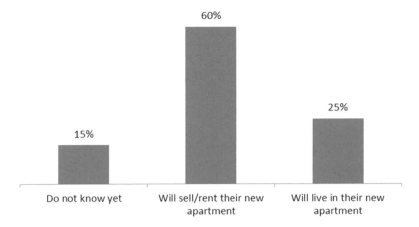

Fig. 10 Resident's intentions regarding the new apartments. (Source: Original survey data (2017))

The decision of apartment owners (70%) who responded that they intend to sell or rent out their new apartment, could be attributed to a simple preference to live elsewhere, but in some cases reflects their inability to afford the new monthly bills: the maintenance charge will grow to 500 NIS (or 100 Euros) (previously 50 NIS-10 Euros), while the bi-monthly municipal taxes will rise to 1200 NIS (compared to 300 NIS on average previously). To assist residents with this new tax burden, the consulting firm has negotiated with the municipality a temporary discount of 30% during the three first years and 20% for the following 2 years, but after 5 years the discount expires.

Some residents reported that they signed the agreement even though their income was not high enough to afford the new monthly payments (Residents' Association Representative 2017: Personal Communication). On the other end of the spectrum, residents refusing to sign the agreement had to face 'terrible social pressures' until they signed. One refuser was sued by neighbors in 2013 (under the 'objector resident' act framework). The court ordered him to sign within 60 days and after that delay to pay 908,000 NIS to compensate the neighbors. The developers also reportedly had a 'secret agreement' with a resident who received 3000 NIS for each household he persuaded to sign (Residents' Association representative 2017: Personal Communication).

Regarding the public social housing tenants' group, the company had no official information on their relocation in the 'Green Park' project or elsewhere, since they left their public social housing in 2014. As social housing tenants they were paying only 300 NIS (about $80) per month for a 65 m^2 apartment and less than 100 NIS in bi-monthly municipal taxes. They were supposed to enter the new apartments, and still did not know how much extra rent they would pay (for their larger apartments), neither for monthly building maintenance (nominally 500 NIS/month) nor for municipality taxes (since they would have a discount on the 1200 NIS basis, instead of 300 NIS as before). The electric bills were also expected to rise due to the larger apartment size, and to the lack of solar water heating systems (in buildings of eight stories and above). However, in 2016 'the urban renewal authorities' (Note 8) (under the planning authorities' supervision) adopted a law aiming to defend the rights of public social housing tenants. According to this law, the difference between the previous and future monthly payments is to be subsidized by the project's developers, so that the tenant will still pay the same amount. In addition, the tenants must be relocated in the new project and the rents cannot be raised. However, it is not stipulated for how long this provision would be in force. Meanwhile, the tenants had no official information from the social housing company on their future conditions.

TEL AVIV 3700: 'Green' and Affordable Housing?

In 2008, together with the other Forum 15 cities, the Tel Aviv-Jaffa municipality signed the Cities for Climate Protection Initiative, which stipulates reducing greenhouse gas emissions by 20% by the year 2020 – and 'green' building according to SI 5281 has been gradually institutionalized in the planning process since January 2013. As long as the standard was strictly voluntary, it was not used to certify any residential building in Tel Aviv – since instead of relying on SI 5281, the municipality had adopted its own guidelines for green building.

In contrast with its past decisions for approving projects, the Tel Aviv municipality has recently declared that 'affordable housing' should be part of all new development plans. According to the Head of the Engineering Department (2017, "personal communication"), this new policy responded to the social protest in the summer of 2011, which had demanded housing affordability. Thus, the TA 3700 project aims to combine 'affordable' and 'green' housing (Fig. 11).

Fig. 11 The location of the TEL AVIV 3700 project

Tel Aviv 3700 is a mixed-use development project covering 1900 dunams of land along a 5 km stretch of the Mediterranean coast – from the Sede Dov Airport (being now removed) in northern Tel Aviv, to the southern border of Herzliyah Pituah.[5] The project is part of a city plan which the Tel Aviv municipality has been developing since 2004, and is to include about 12,000 housing units, office space, hotels, commercial areas, parks, communal areas and a beach promenade.

The development plan was approved in October 2013, and the detailed master plan - in December 2016. To promote the project, the Tel Aviv municipality divided the area into five planning zones, with construction scheduled to proceed from north to south.

The master plan calls building 11,500–13,000 housing units, including 2160 units defined as 'affordable housing' and 1000–4000 'small' apartments. The plan also specifies commercial buildings (147,000 sq.m.), office space (68,000 sq.m.), hotels (60,000 sq.m.) and open spaces totaling 200 dunams.

Along the west side of the development, an ecological beach park is planned, 'preserving and strengthening natural values and making them accessible to the public, to generate a contrast with the active urban neighborhood'. A comprehensive ecological survey was conducted in 2014 by a multidisciplinary team, to ensure the conservation of the cliff running along the beach and determine the right balance between conservation and urban development. The masterplan has been approved by the local, district and national committees, as a condition for approval of each of the five plans.

To integrate 'green' building in the neighborhood planning, the Leadership in Energy and Environmental Design for Neighborhood Development (LEED ND) rating system was selected. It is the most recognized tool for evaluating 'sustainable' neighborhood design in North America (Sharifi and Murayama 2013). The LEED ND rating system, developed by the U.S. Green Building Council (USGBC), incorporates "green building" principles into its criteria (Kibert 2013), rewarding high-density, compact development containing a variety of unit sizes and building types, as well as access to diverse land uses (Fig. 12).

According to the Head of the Engineering Unit, the Israeli "Sustainable neighborhood framework was not advanced enough to be considered. [...] So we adopted the LEED ND and not the British BREAM, as some Israeli consulting firms are accredited to give the LEED ND [...] Thus an environmental consulting firm was hired to help the municipality meet the standard requirements [...] This is the first time in Israel that such a plan has been developed [...] In Tel Aviv we are rich enough to hire the best advisors."

The LEED ND scheme (Fig. 30) incorporates five categories: (1) Smart location and linkage; (2) Neighborhood Pattern and Design; (3) Green Infrastructure and Buildings; (4) Innovation and Design Process; and (5) Regional Priority Credit. The five categories include mandatory and optional credits worth a total of 110 possible points, and a minimum of 40 points must be earned for certification. There are four certification levels according to the accumulated points: Certified (40–49), Silver (50–59), Gold (60–79) and Platinum (80+).

---

[5] An affluent beachfront neighborhood in the western part of the city of Herzliya, Israel, in the Tel Aviv District. It has about 10,000 residents.

Credit	Smart Location & Linkage	27 Possible Points
Prereq 1	Smart Location	Required
Prereq 2	Imperiled Species and Ecological Communities	Required
Prereq 3	Wetland and Waterbody Conservation	Required
Prereq 4	Agricultural Land Conservation	Required
Prereq 5	Floodplain Avoidance	Required
Credit 1	Preferred Location	10
Credit 2	Brownfield Redevelopment	2
Credit 3	Reduced Automobile Dependence	7
Credit 4	Bicycle Network	1
Credit 5	Housing and Jobs Proximity	3
Credit 6	Steep Slope Protection	1
Credit 7	Site Design for Habitat or Wetlands Conservation	1
Credit 8	Restoration of Habitat or Wetlands	1
Credit 9	Conservation Management of Habitat or Wetlands	1
Neighborhood Pattern & Design		**44 Possible Points**
Prereq 1	Walkable Streets	Required
Prereq 2	Compact Development	Required
Prereq 3	Connected and Open Community	Required
Credit 1	Walkable Streets	12
Credit 2	Compact Development	6
Credit 3	Mixed Use Neighborhood Centers	4
Credit 4	Mixed Income Diverse Communities	7
Credit 5	Reduced Parking Footprint	1
Credit 6	Street Network	2
Credit 7	Transit Facilities	1
Credit 8	Transportation Demand Management	2

Fig. 12 The LEED ND rating system – categories, credits and points. (Source: U.S. Green Building Council (2016))

Credit 9	Access to Civic and Public Spaces	1
Credit 10	Access to Recreation Facilities	1
Credit 11	Visitability and Universal Design	1
Credit 12	Community Outreach and Involvement	2
Credit 13	Local Food Production	1
Credit 14	Tree Lined and Shaded Streets	2
Credit 15	Neighborhood Schools	1
Green Infrastructure and Buildings Points		**29 Possible**
Prereq 1	Certified Green Building	Required
Prereq 2	Minimum Building Energy Efficiency	Required
Prereq 3	Minimum Building Water Efficiency	Required
Prereq 4	Construction Activity Pollution Prevention	Required
Credit 1	Certified Green Building	5
Credit 2	Building Energy Efficiency	2
Credit 3	Building Water Efficiency	1
Credit 4	Water Efficient Landscaping	1
Credit 5	Existing Building Use	1
Credit 6	Historic Preservation and Adaptive Reuse	1
Credit 7	Minimized Site Disturbance in Design and Construction	1
Credit 8	Stormwater Management	4
Credit 9	Heat Island Reduction	1
Credit 10	Solar Orientation	1
Credit 11	On-Site Renewable Energy Sources	3
Credit 12	District Heating & Cooling	2
Credit 13	Infrastructure Energy Efficiency	1
Credit 14	Wastewater Management	2
Credit 15	Recycled Content Infrastructure	1

Fig. 12 (continued)

Credit 16 Solid Waste Management Infrastructure	1
Credit 17 Light Pollution Reduction	1
Innovation & design process points	**6 possible**
Credit 1 Innovation and Exemplary Performance	5
Credit 2 LEED Accredited Professional	1
Regional Priority Credit points	**4 possible**

Fig. 12 (continued)

The Tel Aviv 3700 plan addressed the following environmental principles:

- Preserving the existing resources of the site – Proximity to the sea: elaboration of a preservation plan and use of recycled materials for infrastructures.
- Run-off management – the standard is intended to provide a comprehensive solution to manage rainfall and reduce flooding risks (for example: pools/ local reservoirs, etc.).
- Emphasizing community life by encouraging pedestrian activity while reducing motorized traffic – Compactness of the complex, optimal connectivity to public areas and buildings at all levels, encouraging the use of the railways and public transportation, pre-planning a continuous system of cycling paths, shaded public spaces and combating the urban heat island effect.[6]
- Energy savings – management and production of energy at the local level from renewable sources, and use of natural gas facilities. Raising the buildings' energy efficiency rating to A and above and perhaps applying for the SI 5282 certification.[7]

We focused on the residential buildings planned for the complex, to examine whether the goal of building housing which is both 'green' and 'affordable' has been addressed.

The municipality has a declared goal of integrating various kinds of housing in the neighborhood, to attract a heterogeneous population and ensure social mix. For this purpose, the plan includes a portion of small apartments (of about 60 sq.m. each) which will constitute about 40% of the housing stock in both the private and the public sector market. Overall, the average apartment size has been limited to 120 sq.m., to 'avoid the proliferation of huge luxury penthouses' (Municipality of Tel Aviv, Head of the Engineering Department 2017) (Figs. 13 and 14).

---

[6] The urban heat island (UHI) effect is a phenomenon whereby a built-up urban area is warmer than its rural surroundings due to the composition of the urbanized terrain and human activities occurring within it.

[7] Israeli Standard SI 5282 defines criteria for the rating of both residential and office buildings according to energy consumption, which is related to the construction of the structure, as well as to the ambient climate.

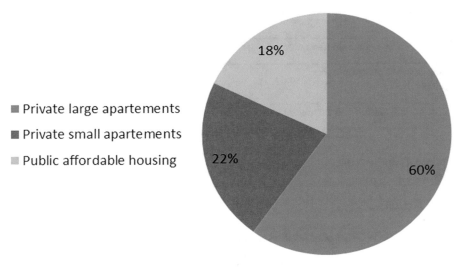

■ Private large apartements

■ Private small apartements

▨ Public affordable housing

Fig. 13 Housing distribution in TA 3700 neighborhood. (Source: Tel Aviv municipality, Engineering Unit (2017))

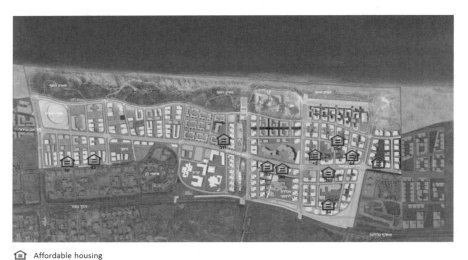

⌂ Affordable housing

Fig. 14 Affordable housing location. (Source: Tel Aviv municipality engineering unit (2017))

The 'social housing' (i.e. smaller apartments) is to be dispersed in the neighborhood, but not mixed with private dwellings at the building scale. They will include no parking spaces, and their planned location along the main road or close to public institutions (such as the Education Campus) is considered less attractive, due to the noise: "We had to be realistic; we could not build the affordable apartments in the most attractive locations, because (a project attempting this) would not happen!" (Head of the Engineering Department 2017).

On the other hand, residents of the neighborhood – including the 'affordable' housing tenants – should be within walking distance of most services. Among the total public rental social housing units (2160), 1360 will be owned by the municipality and 800 by the State. These social units will be funded by private developers receiving the land for free from the State or the municipality, and for the first 20 years will collect rent from the tenants – after which the apartments will be ceded to the municipality, which will collect the rent. It is stipulated that the rent will be 30% lower than the market price, and a given household may stay in the apartment for up to 5 years, to 'enable as many people as possible to enjoy affordable housing' (2017, Municipality of Tel Aviv, Head of the Engineering Department, "personal communication"). Regarding eligibility, households currently owning an apartment cannot apply – but the maximum income has not yet been defined. However, there will be a *minimum* income threshold which stipulates that households for whom the monthly rent represents more than 30% of their monthly income, will be unable to apply for public social housing (2017, Municipality of Tel Aviv, Head of the Engineering Unit, "personal communication"). The 30% level correlates with a reference income based on the 7th socio-economic population cluster (which is the average population cluster in Tel Aviv, whereas North Tel Aviv is mainly populated by the 8th, 9th and 10th clusters). According to the Head of the Engineering Department (2017), the aim is to allow the middle class (7th cluster) into the neighborhood: "This affordable housing is not intended for poor people, it is not public social housing [...] The rents will be around 5000 NIS, so they should not represent more than 30% of the household incomes of the middle class [...] But if the government wants, it can also provide public social housing from its 800 units... it is the duty of the State to build housing for low income people, not ours [...] The only thing we can do is building environmentally certified apartments, while trying to limit speculation. Maybe we are wrong, but at least we are trying [...] The State owns 70% of the land, so they could build public social housing, and some NGOs even try to convince it to do so – but we do not have enough power".

It thus seems that while an unprecedented effort is being made to promote in Tel Aviv 'green' building accessible to a larger cross-section of the population, the 'affordable' housing in the TA 3700 project will be directed to the middle and upper-middle class, thus excluding poorer households. Since the neighborhood aims to receive the LEED ND certification, we could ask whether that framework includes clauses relating to social diversity or affordability for low-income populations, like in the French 'eco-quartier' framework (Machline et al. 2016).

In the LEED ND 'Neighborhood Pattern and Design' category (see Fig.) there are seven optional points dedicated to affordable housing in the sub-category entitled 'Mixed Income Diverse Communities' (formerly known as 'Housing Types and Affordability' – see Fig.). According to Garde (2009), however, the LEED ND certification provides very little incentive or reward for the provision of affordable housing. More recently, Szibbo (2016) found that in 60% of the LEED ND projects in the US, there is no affordable housing (Fig. 15).

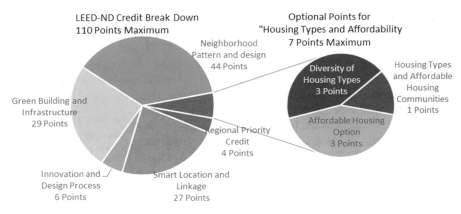

Fig. 15 The LEED ND (v3) credit distribution, related to 'Housing Types and Affordability' (relabeled 'Mixed Income Diverse Communities' in v4). (Source: Szibbo (2016))

Table 4 Table from LEED ND guidelines defining thresholds for affordable housing

Rental dwelling units				For-sale dwelling units				
Priced up to 60% AMI		Priced up to 80% AMI		Priced up to 100% AMI		Priced up to 120% AMI		
Percentage of total rental units	Points	Percentage of total rental units	Points	Percentage of total for-sale units	Points	Percentage of total for-sale units	Points	
5	1	10	1	5	1	8	1	
10	2	15	2	10	2	12	2	
15	3	25	3	15	3	–	–	

Source: LEED N
Source: LEED ND (2016)

The TA 3700 project follows the LEED ND guidelines that stipulate:

Include a proportion of new rental and/or for-sale dwelling units priced for households earning less than the area median income (AMI). Rental units must be maintained at affordable levels for a minimum of 15 years. Existing dwelling units are exempt from requirement calculations. To receive up to a maximum of three points, the project must meet any combination of thresholds in

Table 4 (LEED ND Guidelines 2016).

Thus, while the LEED ND certification includes a small incentive to provide rental units priced to accommodate below-average income, it stops far short of requiring affordable housing. Accordingly, the Tel Aviv municipality can build a LEED ND neighborhood without accommodating low-income populations. Even the initiative to include apartments for the middle class (7th cluster) was not due to LEED ND certification requirements, but rather to a municipal decision following the 2011 social protest over the lack of affordable housing (see chapter "Introduction"):

Before the social protest we planned only 360 affordable units in TA 3700, the other 1000 were to be affordable housing due to smaller size apartments, and thus cheaper (the master plan stipulates at least 20%) […] If we planned a neighborhood at present, we would not do that! (Head of the Engineering Unit, Tel Aviv municipality 2017).

Affordable Green Housing in Tel Aviv?

Both 'green' neighborhood projects in Tel Aviv, Green Park and TA 3700, are essentially residential developments directed to the middle and upper middle class. Pinui-binui is a national project to encourage more 'efficient' land use in urban areas, by replacing medium-density, multi-family residences with high-rise buildings, designed to meet much higher standards.

Following the clearance and redevelopment of Neve Sharett in north Tel Aviv, 154 homes built in the 1960s as public housing have become 447 apartments in six towers (certified 'green'). The site had a high rate of private rentals, whether by initial tenants who became owners, or by investors who purchased the flats intending to redevelop. The program does not create nor preserve affordable housing. Thus, virtually all non-owner tenants (40% of the residents) had to leave the neighborhood when the pinui-binui project was launched. Among the owners, our survey indicates that 70% will be unable to stay in the 'green' neighborhood. In sum, the 'greening' of Neve Sharett has led to gentrification and population displacement.

At the same time, the urban renewal authorities' 2016 law aimed at preventing the public housing tenants from being moved out, will technically allow this low-income population to live in upgraded apartment buildings for an affordable price. If these public dwelling tenants (16 households, or 10% of the total) are really resettled within 'Green Park', their housing units may be considered 'green' and affordable (though this would be an unintended project outcome, and as mentioned, remains uncertain).

The Tel Aviv 3700 master plan was developed over the years 2000–2015, and in the past 2 years, five teams of architects have been working on the detailed planning of the neighborhood's five sub-divisions. One the aims of the overall plan was to avoid single-use zoning (like in the Neve Sharett 'green' neighborhood, where all buildings are residential) – even though residents and developers have both expressed opposition to commercial buildings, which they see as reducing real estate value. In TA 3700 commercial buildings are mandatory along the main streets, and the main guidelines call for mixed uses, high density streets, public transportation development together with pedestrian and cycling paths, a pneumatic waste system, use of natural gas, and a decrease in the number of cars from 2 to 1.5 per apartment.

The neighborhood maintains continuity with the city center through its traffic linkage (along Ibn Gvirol Street) and the apartment buildings will range from 2 and 15 stories tall. The neighborhood's prime location (in northern Tel Aviv, adjoining the seafront) provided an impetus to offer affordable housing, and unlike in the Neve Sharett project, the municipality (which has rights to the land through its partial ownership) did undertake such an effort. The units will belong to the private developer building them, who will not be charged for the land and will collect rent for 20–25 years before the apartments revert to the municipality. A total of 1360 units will be owned by the municipality, and 800 by the State. The units designated as 'affordable' housing are meant for the middle-class (as mentioned, the rental fee will represent 30% of the average net income in the 7th socio-economic cluster) and there are no apartments intended for lower income groups (due to the stipulation that households for whom the monthly rent of around 5000 NIS represents more than 30% of its net income will not be eligible). Apartment rental by a given tenant will be limited to a maximum of 5 years, possibly limiting the extent to which an 'affordable community' can evolve over time. While these buildings are to be submitted for 'green' certification together with the rest of the project (which may dampen their stigmatization as 'affordable housing'), they will be located in less attractive areas (e.g. along the noisy main street).

Finally, to the extent that lower income groups (clusters 6 and below) are only able to rent in the private sector, the TA 3700 initiative will not impede the larger, ongoing trend – whereby these groups are priced out of Tel Aviv.

Peripheral Areas Case Studies

The following two case studies are in Yavneh and Dimona, outside of the Tel Aviv metropolitan area. Both were established as development towns, urban settlements newly built or significantly expanded by the State. A total of 28 such towns were established, mainly in the 1950s, to settle immigrants. Development towns were originally designed for a mixed ethnic population, but it was the newly arrived Mizrahi Jews from low socioeconomic background – mainly those already residing in temporary immigrant camps – who had little option but to stay in the peripheral locations where the camps had been set up.

Most development towns quickly became dominated by low-income Mizrahi immigrant populations, mainly from North Africa (Hasson 1981). The combination of their peripheral location, cultural segregation and economic dependency led the development towns to remain the least developed sector in Israeli Jewish society (Gradus 1984). According to Israel's Central Bureau of Statistics, these towns still form the vast majority of Israel's poor localities. However, the socioeconomic ranking of the development towns closer to the central Tel Aviv region is just below average, while more peripheral towns are in the bottom 20 percent, indicating a strong link between centrality and prosperity. This link is confirmed by a study which ranked all 118 Israeli Jewish urban localities according to their aggregate quality of life indicators (Ministry of Interior 1996). Eighteen of the last 20 ranked localities were development towns, 17 of them in the country's northern or southern peripheries.

We analyzed 'green' building policies and practices in two development towns. The first is Yavneh, whose proximity to Tel Aviv (20 km) made the settlement of middle class households possible since the late 1970s. Thus a 'green' neighborhood development project including SI 5281 buildings was a step in the long gentrification process that began several decades ago. Our second case study, Dimona (in the remote eastern Negev) is still among the poorest Jewish towns. A 'green' neighborhood is being developed in Dimona, and marketed as 'built according to Israeli green standards'. However, the stakeholders involved decided not to seek formal 'green' certification under SI 5281, as they perceived that doing so would threaten the very implementation of the project.

The Case of Yavne: 'Green' Neighborhood or 'Greentrification'?

Yavneh is a city in the Central District of Israel. In March 2016 it numbered 45,059 inhabitants (socio-economic cluster 6). Its population is young: about 36% are in the 0–21 age group, and 64% are below 40.

Yavneh is located 20 km south of Tel Aviv- Jaffa, 15 km northeast of Ashdod, and 7 km east of the Mediterranean. Yavneh was established in 1948 as a transit camp for Jewish immigrants from Arab countries, Iran and Europe. The first neighborhood was built in 1949. In the first phase, its dwellers subsisted on small trade, labor in farming and industry, and some agriculture in their backyards. Living standards were low, housing often substandard, and social welfare cases numerous. Its small population of 1600 in 1953 grew to 10,100 in 1970, and by the mid-1990s to 25,600 – further rising to 31,700 in 2002 and occupying 30 sq. km. Yavneh received the status of urban municipality in 1986.

In the 60s, several enterprises moved from the Tel Aviv area to Yavneh. Its industries include leather, textiles, and metallurgy. With its location near the southern fringe of the Tel Aviv metropolitan area, Yavne was a stagnating backward city until the mid-1970s. Its Mayor, Meir Shitrit, elected in 1974, opted to develop it as a low-density satellite of Tel Aviv. He initiated a project of single-family dwellings for medium-income households, opening a new range of opportunities for Yavneh (Gradus et al. 2006).

The housing project was followed by an upgrade of the school system and attracted a group of army officers and their families to build their houses in the once-depressed immigrant town. The growth relied upon the short commuting distance to Tel Aviv. This rapid transformation was not without problems, however, bringing social polarization and rising land and housing prices (Gradus et al. 2006).

Yavneh eventually became the site of Neot Rabin at the beginning of the 2000s, one of Israel's first 'green neighborhoods.' While another 'green' neighborhood was planned in Kfar Saba in 1996, Yavneh's initiative was the first to include 'green' certified residential buildings (SI 5281, 2005 version). Yavneh does not belong to Forum 15, and thus the standard is not mandatory, even today (Fig. 16).

Fig. 16 'Green' neighborhood location. (Source: Yavneh municipality (2015), Personal communication)

Yavneh's 'green' neighborhood was designed in two stages. The first project of 2900 housing units was approved in 2002, and the second, with 1335 additional units, in 2010.

While the Yavneh project may be thought of as Israel's 'first certified green neighborhood' due the SI 5281 certification of its buildings rather than to overall design, the municipality also employed an environmental consultant, to set general sustainability guidelines.

According to the Head of the Engineering Unit of Yavneh municipality (2017), Personal communication, planning a 'green' neighborhood was a wish of the mayor and "it has been planned and constructed with an emphasis on environmental components [...] There was a need for planned green neighborhoods in Israel, due to problems caused by high-density development, the intensive use of infrastructures and the desire to raise the standards of living, while reducing home and maintenance costs".

According to the Head of the Engineering Unit, the planning of the green neighborhood was carried out in cooperation with the project's architect, the mayor, the municipal engineer, the developer's representatives, and professional consultants dealing with environment, infrastructure, water and waste issues. "The biggest obstacle to green neighborhood construction comes from the larger capital costs (between 2% and 10%) due to the use of expensive technologies and materials [...] These costs fall mainly on developers and on the local authority [...] Meanwhile, those who benefit most are the residents who live in a better quality environment, and in a building where heating and cooling costs are lower (by 30%), thanks to investment in insulation" (Head of the Engineering Unit, Yavneh 2017).

According to the same official: 'green' building is considered to be of higher quality, and in greater demand on the Israeli market, than conventional building.[...] The perception of green neighborhoods as prestigious, causes many local authorities to encourage them, in the hope of attracting families with higher socioeconomic standards, which will also pay more to dwell in such an area.' "Strict standard requirements have been applied at every stage of the construction process; close supervision has been necessary to verify that the developers complied with all neighborhood requirements and environmental regulations. We did not need to check if the SI 5281 standard was efficient, since the Ministry of Environment Protection already did that, and we trust them" (Head of the Engineering Unit, Yavneh 2017).

To improve thermal comfort in buildings, all residential units were designed in accordance with Israeli standards – compulsory (SI 1045) and voluntary (SI 5281) – for thermal insulation of building walls and windows. However, the neighborhood layout was already planned when the project was submitted for certification and the apartment buildings were not oriented properly. Thus, passive cooling and heating are limited, and the buildings received only 55 points of the 2005 standard version, barely enough for qualification (55 points is the minimum for IS 5281 certification).

The last part of the 'green' neighborhood plan was designed in 2010 and includes 1335 residential units on 1606 dunams. There are building rights for 26,950 sq.m. for commerce and offices, about 89 dunams for public buildings, 9.5 for sports and recreational activities and, over 630 for a coastal park forest. The municipality decided that this part of the 'green' neighborhood' will be built according to the LEED ND system, and the residential buildings according to LEED as well – in contrast with other 'green' neighborhoods in Israel, designed to comply with the local green building standard (SI 5281). According to the Head of the Engineering Unit (2017), "the LEED standard is more efficient and better known, which will make it easier to sell the apartments".

As mentioned (see Green neighborhoods in North Tel Aviv), in the LEED ND standard there are several credits encouraging affordable housing but no mandatory requirement. At the same time, 538 housing units in the new section of Neot Rabin will be subsidized by the State. This is due to a government housing program called *Mechir La'Mishtaken*[8] (2016), whereby the Israel Lands Authority and the Ministry of Housing auction off land at a discount to developers willing to guarantee lower-priced apartments to eligible purchasers. Any single or married Israeli over the age of 35 who has not owned an apartment in the last 6 years, is eligible. In other words, about one quarter of the apartments in this LEED ND neighborhood will have a reduced price by about one million NIS (22% below market price).

As of 2016, 89 residential buildings (with a total of 3200 apartments) in the 'green' neighborhood of Neot Rabin had received the SI 5281 certification. Thanks to the project, Yavneh's population, which was 34,000 in 2012, has reached 46,000 –an increase of 35%. A full quarter of Yavneh's habitants currently live in the 'green' neighborhood (2019) (Fig. 17).

To calculate the 'green' premium in Yavneh, we compared the sales price per square meter of apartments in Neot Rabin (the 'green' neighborhood) with those in the adjacent 'Neot Begin' neighborhood –built a few years earlier (in 2004), which also consists of multi-family apartment buildings, but without 'green' certification (See Fig. *). According to this comparison, the 2012 'green' premium in Yavneh is estimated at 13.5%, even though prices in Neot Begin grew by 42% between 2004 and 2010 (according to Tax Authority data). These results indicate that the 'green' neighborhood tends to make surrounding neighborhoods more attractive, driving up real estate prices.

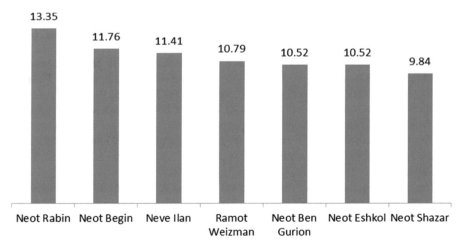

Fig. 17 Average apartment sales price (in thousands NIS/sq.m.) in 2012 in Yavneh's neighborhoods, including Neot Rabin (the 'green' neighborhood). (Source: Yavneh Municipality (2014), Personal Communication)

---

[8]An initiative promoted by MK's Moshe Kahlon and Yoav Galant from the Kulanu party, who have staked their political reputation on increasing housing affordability.

As mentioned, however, one quarter of the apartments in the last sector of Neot Rabin (designed according to LEED ND) will be discounted by 22% due to the *Mechir La'Mishtaken* program. This is in sharp contrast with current trends in the center of the country, as few *Mechir La'Mishtaken* apartments are available in major cities: none in Tel-Aviv, Netanya, Ashdod, or Herzliya. In Jerusalem, about 500 apartments are available in Ramat Shlomo (an ultra-orthodox Jewish neighborhood where prices are already more affordable than in other parts of Jerusalem). Apartments are available for those willing to move to peripheral towns like Rosh Ha'ayin, Charish, or Afula located respectively at a distance of 30, 70, 90 km (with 1700 apartments) – but in most high-demand cities there are none (see http://www.ynet.co.il/articles/0,7340,L-4843203,00.html) .

Thus, while the program does allow those willing to move to the periphery more affordable housing, one may ask: Weren't prices already affordable in Charish or Afula? Are Israelis living in the Tel-Aviv area going to move to Afula, to obtain a 200,000 NIS discount on an apartment?

Comparing this socio-economic data (See Fig.) to the 2008 population census of the Israeli Central Bureau of Statistics (before the building of the 'green' neighborhood), Neot Rabin is a gentrifying neighborhood: (1) Less than 7% of Yavneh residents have an academic degree, versus 50% in the 'green' neighborhood. (2) Only 72% of Yavneh's households have at least one car, while in Neot Rabin, almost half population has two cars or more. (3) Average apartment size is larger in Neot Rabin than in Yavneh. (4) Average apartment price is 13,300 NIS/sq.m. in Neot Rabin versus 10,800 NIS/sq.m. in other Yavneh neighborhoods.

All apartments in Neot Rabin have been built according to the SI 5281 standard (2005 version, 55 points), and received the 'green' label from the SII. We asked the interviewees if they knew that their apartment had 'green' certification, and half the sample (47%) answered that they did not (Figs. 18, 19, and 20).

According to a post-occupancy survey done among 187 residents (see appendix) that moved to the Neot Rabin 'green' neighborhood of Yavneh between 2012 and 2016. The majority (74%) settled in 2013–2015, and most (93%) purchased new

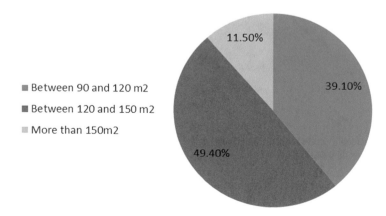

Fig. 18 Apartment sizes in Neot Rabin. (Source: original survey data (2017))

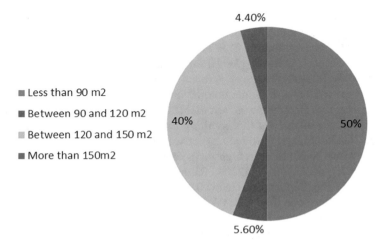

Fig. 19 Previous apartment size of Neot Rabin residents. (Source: original survey data (2017))

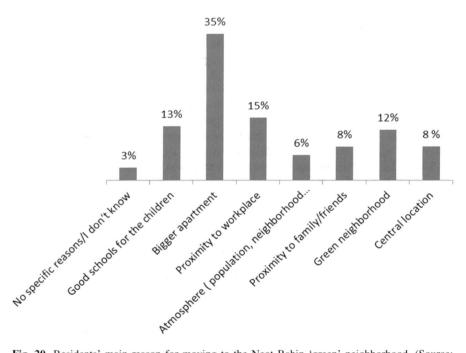

Fig. 20 Residents' main reason for moving to the Neot Rabin 'green' neighborhood. (Source: original survey data (2017))

apartments (5% bought them second-hand). Then, only 12% of the sample declared that they selected that neighborhood because of its 'green' elements. The main motivation was moving into a larger apartment. When we asked residents previously living in a similar sized apartment, only 30% declared that their energy consumption has decreased since they moved to a 'green' apartment.

Meanwhile a computer energy simulation done by the project's environmental consulting firm, has shown that buildings designed to meet the revised (2011) version of the standard may have a 20–25% lower energy demand than conventional buildings. However, a similar computer simulation, conducted for the present study using EnergyUI software (Amir 2017, Personal communication), showed an improvement of 7% compared to a baseline building which meets the minimum requirements of the mandatory SI 1045 standard.

One source of discrepancy in the estimation of certified buildings' energy performance is that so far, most buildings have been built according to the earlier (2005) version of the standard, which does not necessarily require them to be more energy efficient than construction meeting the mandatory standard (SI 1045) for thermal insulation. According to the environmental consulting firm, most buildings in Yavneh's other neighborhoods do not even meet the requirements of SI 1045 (2017, "personal communication"). Regarding water expenses, only 3% claimed that their bills decreased since they moved to the 'green' neighborhood.

According to the developer's environmental consultant (2017, "personal communication"), the residents do not use the air conditioner water collector system to save water, and the solar panels are not even connected to the electricity grid. "The building residents' committee considers that maintenance costs would be too high compared to the water/electricity saving potential, and thus the access to the roof and the water collection system has been locked". After further investigation among building residents (2017, "personal communication"): "We do not use the photovoltaic panels to produce electricity, since none of us is interested in opening a private company (to manage the logistics of a communal system) [...] Regarding the water collector system, the building company provided a pump with insufficient power, and thus the water was not used. It was not profitable due to its electricity demand, and we did not want to invest in a better pump."

Furthermore, a representative of the Residents Council (2017, "personal communication") expressed doubts about the pneumatic recycling system – which aside from eliminating collection vehicles within the neighborhood, "is not greener than a regular waste collection system [...] Wet and dry wastes are not separated, and are sent to the regular dump site and treated as in any other Yavneh neighborhood [...] Our expensive pneumatic waste system is just more esthetic and comfortable".

Regarding the issue of health and well-being, SI 5281 text states that "Tenants in buildings designed correctly are healthier and more relaxed" and the head of the 'green' building unit in the SII declared that the standard "has the potential to reduce workers sickness vacancy and reduce the social insurance budget dedicated to that purpose". Thus, we asked the residents of Neot Rabin if they their working sickness vacancy had noticeably decreased since they moved to environmentally certified apartments. Only 12% answered that it had, while most interviewees complained of an increase in respiratory problems, due to smoke inhalation from surrounding illegal waste burning.

One clearly apparent 'green' aspect in Neot Rabin is its large number of open spaces – and for this reason, higher municipal taxes than in other districts of Yavneh. Comparative data (Yavneh Municipality 2017, Personal Communication) show that Neot Rabin residents have been charged 52 NIS/sq.m., whereas the tax in Yavneh's other neighborhoods ranges between 39 and 49 NIS/sq.m.

Despite its abundance of open space, the Yavneh 'green' neighborhood did not undergo certification by any eco-district standard at the urban planning (as opposed to individual building) level. In countries where it exists, green certification at the neighborhood level tends to emphasize reduced use of private cars as the foremost goal – since 25% of fossil fuel-related CO_2 emissions globally, and over 30% in OECD countries, can be attributed to transportation (Dargay and Gately 1997). For example, in LEED ND, the category 'Neighborhood pattern and design' – dealing with displacements, accessibility, soft transportation development, and local services to reduce the use of private cars – accounts for 44 points and is thus the main requirement for certification. However, the main means of transportation in Yavneh's 'green' neighborhood remains the private car (Fig. 21).

According to the survey sample (Fig. 44), a large portion of Neot Rabin residents are dissatisfied by public transportation. Due to lack of local retail shops, public services (such as schools), and sources of employment (with no tertiary buildings) within the core of the neighborhood, most daily travel has to be by car. This can explain why more than 60% of households in the sample own (or have the use of) at least two cars. Thus, Neot Rabin would be ineligible for LEED ND, or similar 'green neighborhood,' certification.

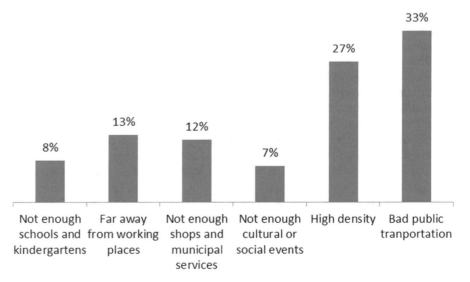

Fig. 21 Responses to the question "What do you dislike in your neighborhood?" (Source: Original survey data (2017))

The Case of Dimona

Dimona is located in the Negev desert, 36 km to the south of Beer-Sheva and 35 km west of the Dead Sea, above the Arava valley in the Southern District of Israel. In 2015 its population was 33,258.

Dimona was one of the ten development towns created in the Negev in the 1950s under the leadership of Israel's first Prime Minister, David Ben-Gurion. Dimona was conceived in 1953, and settled in 1955, mostly by new immigrants from North Africa, but with a sizable Indian Jewish community (about 7500 residents). Its population often lacked the education and skills necessary for higher-level employment. Furthermore, veteran Israelis were typically absent from the new towns, preferring the coastal cities or kibbutzim (communal agricultural settlements). This limited town development (Comay and Kirschenbaum 1973). Housing conditions were rudimentary, due to financial hardship of the immigrants and of the State. Temporary immigrant camps were constructed initially, and in their place, government housing corporations typically constructed multi-story apartment blocks. In 1980, public housing accounted for 95% of Dimona's housing stock (CBS 1981).

With its relative proximity to Beer-Sheva, Dimona played an important role in the development of the Negev region, but its distance from the country center was a serious disadvantage. Unemployment was gradually reduced, thanks to government-assisted industrialization (Silberberg 1973). The labor-intensive textile industry was seen by policy makers as cheaper in terms of investment, and suitable for Negev development towns, and by the early 1970s it employed 40% of the region's industrial manpower (Gradus and Einy 1981). Many plants closed in the 1980s, as taxes on textile imports were reduced. At the same time, development of the natural resources of the Negev, notably the potash deposits of the Dead Sea (and later the phosphate rock of the northern and central Negev) progressed. About a third of the city's population works in industrial plants (chemical plants like the Dead Sea Works, and high-tech companies). In spite of a gradual decrease in the 1980s, the city's population grew once again with the large-scale immigration from the former Soviet Union in the 1990s.

Many workers were made redundant in recent years, generating a 10% unemployment rate. However, Dimona took part in Israel's solar energy transformation, as Luz II, Ltd. constructed a large experimental thermal solar array at the Rotem Industrial Complex, outside of the city. With dozens of mirrors focusing the sun's rays on a tower, that in turn heats water to create steam and generate electricity, the installation is being billed as the 'highest performance, lowest cost thermal solar system in the world,' and the company hopes to implement the technology in new solar plants, in California.

In 2008, a master plan for Dimona's new 'Shahar' neighborhood was approved. The project is located at a 1200 dunam site in the northeast of the city, on a former municipal garbage dump. The program includes 3412 residential units (about 15,000 residents by 2025), five schools, and ten kindergartens, at a total development cost of NIS 500 million. According to the deputy mayor of Dimona (2017), Personal Communication the aim is a to attract "a more affluent population from the center of the country, but also to host families from military units that moved to the region [...] and to retain Dimona's youngsters [...] The Ministry of Housing entrusted infrastructure construction to a public developer, assuming that no private company would invest in the 'unattractive peripheries'" (Head of the Development Company 2017).

The land was divided into lots and the developer published a call for private tenders to take charge of building and marketing the housing units. The plan includes housing construction as follows:

- 547 units for low apartment buildings (4 units per dunam)
- 1977 units in apartments building
- 119 cottages
- 364 garden and roof apartments
- 209 private houses
- 200 housing units for the elderly

So far, each of the 200 lots for private houses was leased from the Israeli Land Authority[9] for 250,000 NIS (including infrastructure development). According to the Demographic Growth Unit of the Dimona municipality (2017), about 70% of the lots have been purchased by Dimona's younger generation. The next stage is to include 584 housing units in apartment buildings, divided into six lots. The Ministry of Housing decided that they will be built under the *'Mehir la'Mishtaken'* framework, thus priced below market-price. The Ministry stipulated that the selling price will be between 5300 and 6200 NIS/sq.m. for apartments of between 60 and 150 sq.m. According to program regulations, half the apartments are intended for Dimona's residents and the other half for outsiders.

The engineering unit of the Dimona municipality was not involved in planning the neighborhood, but may have affected local planning committee decisions. The Ministry of the Interior employed an architect to design the master plan and the public developer established building requirements in the call for tenders.

The inclusion of 'sustainable' elements in planning originated in a government decision (in 2011) committing public developers to 'sustainable development strategies' and thus employing an environmental consultant to define 'green' objectives (Head of the Development Company 2017: Personal communication). The decision to market the project as a 'green neighborhood,' according to the consultant, was the developer's idea: "It was thought that the neighborhood's location on a reclaimed waste dump site, and the use of recycled waste in the building of infrastructure, makes it 'green' [...] but maybe it should not be called a 'green neighborhood' as no green standard is involved" (Environmental Consultant 2017: Personal communication).

The use of the Israeli green building standard (SI 5281), advised by the consultant, was rejected by the developer and the local planning committee as "it would raise building prices and no contractor would invest in Dimona [...] We are not in

---

[9] The Israel Land Authority (ILA) is the government agency responsible for managing this land which comprises 4,820,500 acres (19,508,000 dunams). If the land is state owned (as is 90% of Israel's land).

Tel Aviv, we do not sell the apartments for two million NIS, especially for *Mehir la'Mishtaken* where the State caps prices [...] When a developer's profit is 40,000 NIS per apartment, he is not going to pay 30,000 NIS to the Israeli Standards Institute and 10,000 NIS to a 'green' building consultant, to get the certification' (Local Planning Committee Member 2017, Personal communication).

According to the environmental consultant, the only remaining option for including 'green' building was to integrate elements 'that would not raise building costs'. Thus, the consultant convinced the developer to build according to the mandatory standard for thermal insulation (SI 1045) 'even though the municipal engineering unit was unable to verify its implementation' (the public developer paid a private consultant to ascertain compliance). The consultant declared that the buildings would reach between 35 and 40 points if they were submitted for the SI 5281 standard (less than the 55 points required for certification). Since no 'green' standard is involved in the neighborhood or building design, one may ask why this project is marketed as 'green'.

According to the building company sales department, the apartments 'have been built according to the Israeli green building standards'. In response to the question of what is 'green' in the design of the apartments, it was stated that the project employs 'double glazing and a buried electricity system' (2017, Building company, sales department representative, "personal communication").

The exploitation of environmental issues is not new to the advertising industry, but until recently, it was mostly limited to ads by energy companies (Donahue 2004). In the last decade, however, 'green' advertisements have multiplied (Catenaccio 2009). Gillespie (2008) attributes this to the considerable increase in public awareness of ecological issues, such as land conservation, recycling, and energy/water consumption (Donahue 2004). According to Gillespie (2008), developers and building companies have found it beneficial to tell potential consumers that they were "going green," even when they were not. The result has been labeled "greenwash," defined by Gillespie (2008) as "advertising or marketing misleading the public by stressing the supposed environmental credentials of a person, company or product when these are unsubstantiated or irrelevant".

In Israel, more and more building projects, residential complexes, and neighborhoods are being advertised as 'green'. These ads are often accompanied by a "green" descriptor, as in 'Green Yavneh', or 'My home in Green Kfar-Saba, and building projects commonly have green names, such as 'EcoTower' and, 'Green Hill'.

An increasing number of cities claim to be 'green' and keeping their 'green promises' (which in Hebrew denotes 'sustainable'), as advertised by the cities of Kfar Saba and Netanya. The city of Hod Ha'sharon boasts its' winning the title of 'Green City'; its slogan is 'A green community town'; a building company markets the prestigious 'Avisror Heights: overseeing a breathtaking view and enjoying a lovely breeze', in 'Green Yavneh'. In these ads, cities promise cleanliness, park development, rivers, green building practices, and bicycle trails.

Meanwhile in the Negev periphery, in low socio-economic status towns like Dimona, the 'official green status' of the SI 5281 standard seems unaffordable – and requiring it could preclude project implementation. So far, only 30 households (since April 2017) have settled in the first completed Shahar neighborhood project, in two-story apartment buildings (not defined as affordable housing). In addition, 200 lots for private houses have been leased (for 250,000 NIS). The Israeli Lands Authority stated that half the lots were meant for Dimona residents and the other half for outsiders. However, according to the Demographic Growth Unit of the Dimona, (2017), outside buyers seldom build a house in the Shahar neighborhood, and half the lots meant for them, have been re-sold to Dimona residents for about 500,000 NIS on average.

We collected socio-economic data on current residents of the Shahar neighborhood (30 households, see Figs. 22, 23, and 24).

All apartments are of 120 sq.m. (four rooms), and 100% of the respondents declared that they save no electricity or water, even though they live in the 'green' neighborhood. The main reasons for moving were getting a larger apartment (50%), living among a 'better population' (40%) and for 10% of the sample 'dwelling in a

Fig. 22 Household size (number of residents) in the Shahar neighborhood. (Source: Original survey data (2017))

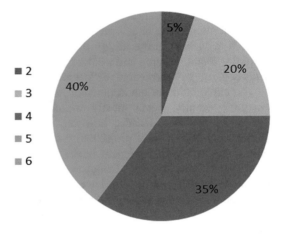

Fig. 23 Children between 0–18 years per household in the Shahar neighborhood. (Source: Original survey data (2017))

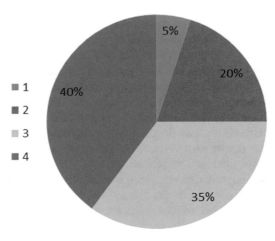

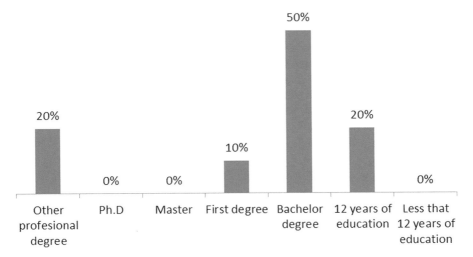

Fig. 24 Educational level in the Shahar neighborhood. (Source: Original survey data (2017))

green neighborhood'. According to the Demographic Growth Unit (2017, "personal communication"), only five households (out of 200) that purchased a lot attended the 'green' building workshop organized by a professional consultant, and the main reason for this lack of interest toward 'green' building concepts was the belief that they would drive up construction costs.

Half the sample own (or have the use) of two cars, and the other half have one. At this early stage (with only 30 apartments lived in) there is no public transportation, no school and no shops, services or work places within walking distance of the neighborhood.

Comparing the Shahar residents' socio-economic data with those of Dimona overall in 2008, we can see that the former evince gentrifying characteristics: (1) In 2008, 10% of Dimona's residents had less than 8 years of education, while in the Shahar neighborhood, 50% had a bachelor degree; (2) More than 50% of Dimona residents lived in apartments of less than three rooms, versus four rooms on average in the Shahar neighborhood; (3) Less than 50% of Dimona residents owned a car, while in the Shahar neighborhood all households have one; (4) Higher apartment prices: compared to another new building project in Dimona, prices are higher by 7% in the Shahar neighborhood.

Dimona's 'green' neighborhood illustrates that in Israel remote, low socio-economic localities have no certified 'green' apartments. However, the 'green' premium still exists, and the neighborhood hosts local middle-class residents seeking a housing upgrade.

In both Yavneh and Dimona, stakeholders have shown an interest in developing 'green' building and 'green' neighborhoods, to attract a wealthier population. In both cases, developments marketed as 'green' offer higher apartment prices without reducing costs (i.e. for electricity or water). While 'green' marketing is omnipresent, only in Yavneh have the developers and the municipality considered building according to the SI 5281 standard even it costs slightly more. In Dimona, while the neighborhood is marketed as meeting 'green building standards,' no apartments have been built according to recognized standards, out of fear that doing so would prevent project implementation. Meanwhile, compared to another new building project in Dimona, prices are higher by 7% in the Shahar neighborhood. At the same time, due to 'green' oriented planning, the apartments have respected the mandatory insulation standard SI 1045, an important improvement for Dimona where most construction fails to respect it.

Conclusion

We conclude is that in centrally located and economically strong municipalities this involves green certification, while in peripheral locations such certification is not implemented – and the 'green' label is mainly used to attract local residents who can afford a housing upgrade. In the most attractive locations, where gentrification is already occurring with housing refurbishment, developers hardly need 'green' certification, and are attracting the well-to-do without it. 'Green' building is notably used as a gentrification tool in undervalued areas of wealthy cities, rather than in those which are seen as attractive by themselves (such as scenic or seaside).

Green Building in Europe

Following publication of the Brundtland report[1] in 1987, "sustainable development" has become a widely accepted conceptual framework for urban policy and development, providing context for a considerable literature on planning, architecture and urban design (Williams et al. 2000). Debates pitting against nature preservation against development interests were no longer limited to environmental aspects, but began to incorporate economic and social dimensions (Lele 1991) – based on the notion that "environmental quality and economic development are interdependent, and in the long term, mutually reinforcing" (Dampier 1982). More recently, the focus of sustainability on controlling change and growth has been broadened to accommodate the concept of resilience, or the capacity of systems to adapt to changes and recover from disturbance (Ahern 2011).

While the social dimension of sustainability raises questions of social justice, there is no agreement on answers (Hopwood et al. 2005) – despite a recent European policy focus on 'sustainable communities' and social cohesion. The 'Bristol Accord[2]' (2005) spells out a common European approach to 'sustainable communities' signed up to by EU member states, building on previous EU initiatives, including the Aalborg Charter[3] and Agenda 21.[4] The concept of social equity within

---

[1] The United Nations World Commission on Environment and Development (WCED) in its 1987 report Our Common Future defines sustainable development: "Development that meets the needs of the present, without compromising the ability of future generations to meet their own needs."

[2] The "Bristol Accord", sets out (1) eight characteristics of a sustainable community; and (i2) contains an agreement to compile good practice case studies that demonstrate sustainable communities' characteristics to an agreed template.

[3] The Aalborg Charter (1994) is an urban environment sustainability initiative approved by the participants of the first European Conference on Sustainable Cities & Towns in Aalborg, Denmark.

[4] Agenda 21 is a voluntarily implemented action plan of the United Nations with regard to sustainable development. It is a product of the UN Conference on Environment and Development (UNCED) held in Rio de Janeiro, Brazil, in 1992.

© The Author(s), under exclusive license to Springer Nature Switzerland AG 2020
E. Machline et al., *Green Neighbourhoods and Eco-gentrification*,
SpringerBriefs in Environmental Science,
https://doi.org/10.1007/978-3-030-38036-6_4

definitions of sustainable development (Hopwood et al. 2005; Chiu 2002) has focused on meeting the needs of future as well as present generations (WCED 1987; Holden and Linnerud 2007) redressing inequities of outcome (Haughton 1999). An equitable society is one with no 'exclusionary' or discriminatory practices, hindering individuals from participating economically, socially and politically (Pierson 2002).

Sweden and the other Nordic nations provide strong evidence that economic prosperity can be combined with social equality and environmental responsibility. In terms of income distribution, they have much higher levels of equality than Australia, Britain, New Zealand and Canada – and nearly twice that of the United States (Andrews et al. 2011). In Nordic countries, private conservation bodies and nature conservation legislation were initiated relatively early - between 1909 (Sweden) and 1923 (Finland) (Koester 1980). All the Nordic countries consider environmental policy a key national interest that they strive to endorse at all levels of international negotiations (Tunkrova 2008).

Social equity is linked to the built environment, either through the provision of services facilities, or by the means of accessing them (e.g. public transportation). Additional criteria include access to decent housing, as measured by its physical condition, but also by the services provided by the relevant housing association/ local authority. Furthermore, problems of housing affordability (and tenure) may prohibit residents from living in or moving out of various neighborhoods and areas. Prompted by an entrepreneurial mode of governance, sustainability has increasingly become a buzzword for urban environmental and political governance (Keil 2005).

In France a national funding system subsidizes public housing construction (the SRU law[5]), and more than half the environmentally certified apartments are defined as public social housing. Additionally, as an EU country, France benefits from European funds earmarked for improving the energy performance of public social housing, as part of the European Energy Performance of Buildings Directive (EPBD). One way to implement the EPBD was establishing a national framework for eco-districts, whereby municipalities receive budgets from the European Union (through the State) subsidizing the construction of 'green' and affordable apartments. The *Eco-quartier* framework places a high priority on encouraging social diversity, and proposing measures to promote social mix is a prerequisite for project approval. According to the Head of the Sustainable Planning Unit – Ministry of Housing, the Equality of Territories and Rurality (2015): "The social aspect is the French peculiarity; a project without social mix, not taking social issues into account, cannot be an eco-quartier. In any case, it will not be an eco-quartier as perceived by the Ministry [...] it is incompatible with how we consider sustainable development in France! [...] The aim is to promote the idea of social mix. We want to give all people the opportunity to live in an eco-quartier. The aim is not to sell the apartments at a higher price". De Chastenet et al. (2016) argue that "we are far from

---

[5] The SRU (Urban Solidarity and Town Planning Renewal) Act, 2000: Stipulates that every municipality of more than 3500 residents must include at least 20% of social housing.

the edge cities fully funded by private groups aiming to produce luxury neighborhoods for urban gentrification". According to program requirements (until 2015), the local authority had to verify "whether housing location and prices are appropriate to household income" in the area.[6]

Following the French mandate for social mix and housing affordability, eco-quartiers are required to have at least 40% public social housing. However, this does not specify the required shares of subsidized housing to be provided for the low, middle, and upper-middle class. Thus, decision makers manipulate funding methods, with the status of the accommodation and the size of apartments, while attempting to generate some social diversity (some apartments are rental social housing). Furthermore, regulations specify that municipalities in deficit must include social housing construction in urban development projects. Thus, more than half of the 'green' certified (BBC) apartments are social housing.

In Denmark and Sweden there is a great deal of municipally owned rental housing. In Denmark, non-profit rental housing, about 20% of all housing, is owned by tenant associations. These dwellings have been built with support from the state and municipalities. Special legislation governs non-profit housing. One requirement for public support can be that some dwellings are allocated based on social criteria. Non-profit housing is not owned by a municipality but by a non-profit housing organization whose sections own the dwellings. Each section has a board of directors, all of them residents. There are over 700 housing organizations in Denmark with more than 7000 member sections. The municipality has supervisory responsibility and has to approve their important decisions.

In Sweden, non-profit housing companies have existed since the 1930s. The great expansion came in 1947 when the provision of housing became a municipal task, mandated by legislation. State loans and subsidies were introduced at the same time. Today, most non-profit housing companies are wholly owned by municipalities. In Sweden's approximately 290 municipalities there are some 300 publicly owned housing companies, owning together about 900,000 dwellings. This represents about 23% of the total housing stock. In contrast to publicly supported rental housing in Denmark, in Sweden a cost-price principle applies to each company and there is no cost redistribution between companies in different municipalities. Municipalities unable to cope with their financial commitments for housing can, however, receive state support. Formally, there is no social rental housing in Sweden but the municipalities may still rent or purchase flats and make them available to needy people. This is essential, as the right to social assistance, mandated by the Social Services Act, includes a right to housing (Elsinga and Lind 2013).

Even though Nordic cities and regions are internationally known for social cohesion and relative social equality, social polarization and fragmentation are seen as severe challenges for the Nordic city-regions. The capital city-regions of Norway, Sweden, Denmark and Finland all exhibit socio-economic segregation.

---

[6]Translated from French: "Les prix des logements et leur localisation sont-ils adaptés aux capacités financières des profils des ménages?"

Vesterbro Eco-district, Copenhagen

In the early 1990s, Vesterbro was the poorest district in Copenhagen. Following its refurbishment into an eco-district in 1994, it has become one of the city's most attractive and expensive neighborhoods.

A former industrial area of Copenhagen, Vesterbro has undergone rapid urbanization since the 1850s. To meet housing needs, this residential neighborhood was rapidly built between 1850 and 1920. Its unhealthy and poor-quality buildings deteriorated throughout the twentieth century. Prior to the rehabilitation of Vesterbro in the 1990s, its dwellers lived in sub-standard conditions: 64% of the apartments had no central heating, nor hot water supply, 71% had no bathrooms and 11% no toilets (Sanchez 2018). Unemployment was around 20% and the crime rate was well above the national average.

The renovation plan, launched by the municipality, aimed to eradicate drugs and prostitution and improve the living conditions of the residents, while respecting the principles of sustainable development, including the social, environmental and economic aspects.

The rehabilitation of Vesterbro has been highly publicized in Denmark and abroad, because of its scale and its technical and social innovations. Indeed, Vesterbro is publicized through the national and local press as a real ecological laboratory in an urban environment, and the project is often presented as a model eco-district. In particular, two blocks of buildings named Hedebygade and Hestestalden have attracted visitors from all over the world, curious to discover innovative technical installations in terms of energy production and low consumption, rainwater harvesting, waste recycling and green space development. Informal guided tours take place regularly. The municipality of Copenhagen, which launched the urban renewal project, selected two semi-private development companies to implement it. The project, completed in 2004, includes 7730 apartments on 950 dunams.

The environmental design approach applies to both buildings and green spaces. First, the aim was to use 'green' materials for buildings. Thus, the municipality in conjunction with the State and the European Union funded two pilot projects: the aforementioned 'Hedebygade and Hestestalden'. While most buildings in Vesterbro have benefited from the same facilities, the two projects have special status, as their funding and monitoring, have allowed more efficient environmental and ecological development. Solar panels and a new high-performance ventilation system were installed, rainwater is collected for toilets, buildings are better insulated, facades of buildings are vegetated, and wastes are sorted into 12 different categories. However, due to a lack of maintenance, the equipment suffers from malfunctions (such as waterpipe leakage and degradation of solar panels).

Increasing the quality and scope of green spaces was one of the aims of neighborhood rehabilitation. If today Vesterbro is the district of Copenhagen that has the fewest square meters of public green space per resident (2 m²), special attention has been given to open spaces, vegetated or not, public and private. The three main parks in the area (Saxopark, Enghaveparken, and Skydebanen) have been refurbished and a former tram line has been converted into a green promenade more than 1 km long (Sønder Boulevard). Moreover, to deal with the shortage of public green spaces, the inner courtyards of most building blocks have been turned into collective green spaces: concrete has been replaced by plants and grasses. These developments have helped to transform Vesterbro and turn it into a showcase of sustainable development.

The second aim of the urban renewal plan was to make Vesterbro a "socially sustainable" neighborhood attractive in the long term (Larsen and Lund Hansen 2008). Even though social mix and integration were considered in the plan, gentrification could not be avoided. The price of land has increased by 80% in 15 years, which has necessarily had repercussions on the social distribution of the neighborhood and has forced the poorest populations to leave. Still, many social structures, such as the Cafe Klare (a social center for homeless women), Settlement (an association fighting for labor market integration of former drug addicts), and Maendenes Hjem (a health and social services center for homeless, drug addicted, and vulnerable men) are located in the heart of the district, which encourages homeless people and drug addicts to visit it on a daily basis.

Social interaction between residents is promoted in the neighborhood, but also inside blocks of buildings, through the development of public and community spaces. For example, throughout Sønder Boulevard, benches, children's playgrounds, bike lanes, and football and basketball courts have been developed, to attract different types of users. In most blocks of buildings, laundries (functioning through a system of rainwater collection) as well as common rooms were made available to residents to encourage meetings between neighbors, meals or parties, cultural activities and sport classes. Community gardens, play areas, picnic tables and shared open space also contribute to social interaction.

Many jobs have been created: Istedgade, a central street of the eco-district, is one of the most important commercial quarters of the capital. Several eco-businesses, as well as many restaurants and coffee shops, have located there – like Løs Market, Denmark's first organic store that does not package its products.

Hammarby Eco-district, Sweden

Developed in 1991, first as a prototype of an eco-responsible sports village to support Sweden's application to host the Olympic Games, Hammarby Sjöstad's project was a first of its kind. The Stockholm municipality, which ran the project, made several calls for tenders for developers, public and private. The neighborhood had to be built in accordance with the *dubbelt så bra* strategy: literally, it had to be built "doubly better" than what was done at the time, in terms of both results (including 'green' technologies) and construction (consuming half the energy of traditional sites). The city of Stockholm also clearly stated its desire to create an eco-district to serve as a model for other sustainable projects. In 1996, the municipality drafted a set of environmental criteria to be met, presented in a report of the China Development Bank in 2015. Work began in the late 1990s and was to be completed by 2015, but the deadline was extended until 2020.

A Sustainable Project Where Social Aspects Were Neglected

In France, as part of the Eco-quartier label, the social component focuses specifically on social mix, with the aim of offering a living environment that promotes solidarity and diversity (Boutaud 2009). In Sweden, the district of Hammarby Sjöstad belongs to a wave of earlier eco-districts (1990s), still very technical, when sustainable development projects were more focused on green technologies than social concerns.

Located south of the main island of Stockholm (Södermalm), the district is in the southern fringe of the municipality, between its administrative boundaries and its southern suburbs, such as Nacka. Formerly economically favored thanks to the digging of the canal between Lake Hammarby and the Baltic Sea, the effects of European competition in the 1950s and 1970s led to the area's abandonment by large industrial groups. Informal activities, such as garages or nightclubs, began to settle in the neighborhood (Canelli 2017). It became an industrial wasteland and was considered dangerous by residents. With the creation of the Metropolitan Region Mälar in the early 1990s, which positioned Stockholm as a showcase for Swedish development, the city decided to rehabilitate that space. Two main motivations have been identified: to offer a pleasant living environment in an area heavily affected by automobile and industrial pollution, and to position itself as a candidate for the 2004 Summer Olympic Games (eventually won by Athens). The city bought the land which had remained private and proceeded to evict illegal residents. The Hammarby project would then be based on two elements defining it socially: on the one hand, housing prices rose due to the cost of cleaning up the area, especially the lake, and on the other hand the neighborhood was unable to accommodate its initial population.

Double-Edged Results

When interviewing residents, three main elements were considered to find out whether there was a real neighborhood life:

1. The presence of places of sociability, conductive to interaction between users. These are usually outdoor spaces, such as walkways or children's parks, located between buildings and facilitating exchanges. In addition, the local community

seems to exist through the neighborhood Facebook group, on which residents are very active: they offer collective activities, complain about nuisances, report lost objects or unusual events, etc. It is also the means used by residents to express pride in their living space.

2. The connection of the neighborhood to the city. This is not an ecological gated community, closed on all sides and only accessible to residents. The area is completely open to the rest of the city of Stockholm and has no fences, even at the buildings edge. Everyone is free to walk through the open spaces around the buildings, and to cross the entire neighborhood through the small courtyards. This openness is one of the neighborhood's requirements. When some condominiums proposed to install grids to limit passage, it generated resident discontent and thus they were not built. In terms of mobility, the area is well served by the subway, which provides rapid access to the city center, the tramway and several bus lines crossing the entire neighborhood, and the free ferry connecting Hammarby to Södermalm, allowing pedestrian and cycling access to the city.

3. Intergenerational mix. Due to the history of the neighborhood, the mix of generations is still very much present. Originally conceived for older age groups because of their greater financial capacity, the neighborhood has in fact attracted families of all generations from the beginning. The municipality had to react quickly, to build the school infrastructure needed to accommodate families with children, which had not been considered initially. The district has reserved some buildings for older residents (the real estate developers having imposed a minimum age of 50–55 years). This is the case, for instance, in the center of the district, at Sickla Udde. This area is one of the least exposed to noise and offers the most pleasant living environment. Prices are higher. Except for this, the neighborhood is a real intergenerational space.

A Closed Community

Interviews with residents reveal that the neighborhood is socially homogeneous and includes no social housing. The resident income distribution confirms the absence of social mix, as prices target only a certain part of the population. Owners are the majority in the real estate park. The latter were asked about the price per m^2 of their properties, in addition to an analysis of about 20 real estate listings: the dispersion oscillates between SEK 50,000 and SEK 80,000 (Swedish kronor, around 5000–8000 euros per m^2). Given that the average Swedish salary is 21,900 SEK (Statistics Sweden 2018), around 2000 euros, and considering the cost of living in the Stockholm area, prices are only accessible to the wealthiest part of the population. The real estate supply in the neighborhood is also quite uniform: most apartments are the same size (more than 80 m^2), arranged in the same way, with at least two bedrooms, a large living room, large windows to enjoy the sun and balconies to watch the surroundings.

A Specific Social Approach?

Most residents seem to regret the lack of social diversity and neighborhood life. This is an interesting prospect to be explored for improvement, although it is paradoxical: indeed, the inhabitants of the district deplore social homogeneity, whereas they belong to the population that generates and sustains it, perhaps unintentionally. If this desire for greater diversity is expressed by residents, it would be interesting for the municipality of Stockholm to consider public spaces where people could express their ideas, not only to deal with pragmatic environmental aspects but also for opening the neighborhood to more social diversity.

Moreover, such a suggestion would imply that social diversity is the ideal situation. Part of the scientific literature has shown that this is not necessarily so. On the one hand, in Hammarby, the neighborhood is not perceived as a golden enclave within the city since its dwellings are average in price. On the other hand, it is possible the model proposed by Hammarby may correspond to a particular vision of sustainability, whereby equality among social groups is an important concern. Such an eco-neighborhood might work in Stockholm, but not necessarily elsewhere. Nevertheless, because of the evolution of its socio-demographic structure (population growth and growing social inequalities), the city may have to rethink this social approach.

These two Scandinavian eco-districts were publicly supported in former substandard neighborhoods of the capitals.

The Scandinavian countries have a social housing policy but 'green' building is no part of it – highlighting the specificity of the French policy (promoting 'green' and affordable housing), even among the most socially oriented European countries.

In large cities like Stockholm and Copenhagen, eco-districts generate gentrification.

The French Case Studies

The surge of environmental awareness and regulation has created 'green' markets, especially in real estate development, leading to the "eco-gentrification" of some neighborhoods (Krueger and Savage 2007). Urban renewal, and projects marketed as "sustainable" neighborhoods, can in fact foster social polarization. One goal of such projects is improving the quality of life in a particular location, generating spaces more attractive than the rest of the city. If located in an already attractive (i.e. expensive) area, such improvement may paradoxically perpetuate discrimination in access to "sustainable" housing.

In France, the State and local authorities encourage projects that are promoted as sustainable neighborhoods, and include 'green' certified buildings. This approach is consistent with the French culture of urban planning, as public authorities are the central actors in urban projects (Souami 2009). In contrast, well-known sustainable neighborhoods in Europe (e.g. Malmo in Sweden, BedZed in England, and Vauban in Germany) have started from private initiatives by dwellers, property developers or associations. However, a process of innovation has emerged with the development of the "Grenelle"1 and 2 (in 2008 and 2010, respectively)[1] environmental initiatives, launched by debates organized by the Ministry of the Environment with associations, researchers, and professionals aiming to establish a coherent strategy for sustainable development.

One prominent element is the "Sustainable City Plan", which contains calls for neighborhood development projects known as "Eco-quartiers" (eco-districts). In promoting eco-quartiers, the ministry has emphasized that the French approach to sustainable neighborhoods does not neglect the social dimension, which is conspicuously absent in other countries. In the French conception, these neighborhoods

---

[1] The environmental conference (Grenelle de l'Environnement) launched by the presidency emphasized the need to reduce energy consumption in buildings and aimed to reduce it to 50–80 kWh/m² (i.e. one quarter of present consumption) over the following 10 years (Tuot 2007).

© The Author(s), under exclusive license to Springer Nature Switzerland AG 2020
E. Machline et al., *Green Neighbourhoods and Eco-gentrification*,
SpringerBriefs in Environmental Science,
https://doi.org/10.1007/978-3-030-38036-6_5

should be mixed, or socially diverse. As explicitly stated, the issue is to avoid the creation of "ecological enclaves for the upper middle class" (Charlot-Valdieu and Outrequin 2009).

'Green' Value in France

The European Union has supported the improvement of building energy performance with a range of legislative and funding mechanisms and instruments. A key part of this legislation is the Energy Performance of Buildings Directive (EPBD). The EPBD was approved on 16 December 2002 and brought into force on 4 January 2003. Its principal objective is to promote the improvement of the energy performance of buildings within the EU through cost-effective measures.

In France, the Ministry of Ecology, Sustainable Development and Energy is responsible for implementing the EPBD. Following a decision in parliament, the French Government published the program's governing law, defining the scope of the energy policy, in July 2005. The execution orders are the government's responsibility. The Energy Performance Certificate (EPC) is called "diagnostic de performance Energetique" (DPE).[2]

The EPC was introduced in France in 2006. As it must be presented when selling or renting a property, it forms part of the real estate diagnosis file, which defines the energy consumption of the dwelling or building, its equivalent greenhouse gas emissions, a breakdown of this consumption for heating, cooling and domestic hot water expressed in final and primary energy, and the corresponding annual costs (see Appendices A). The energy label grades buildings on an energy consumption scale ranging from A (low energy consumption, high efficiency) to G (high energy consumption, poor efficiency).

---

[2] Ministere de l'Ecologie, du Develppement Durable et de l'Energie website: http://www.developpement-durable.gouv.fr/-Diagnostic-de-Performance,855-.html

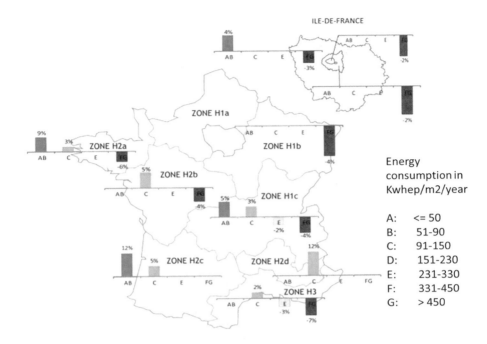

Price increases for apartments receiving higher energy tags (in % comparing to the price of a similar apartment with energy tag D in the various climatic zones). (Map: Dinamic (2014); Source: Notarial Database BIEN)

In France, there is a correlation between the energy saving potential and sale price: apartments in multi-dwelling buildings that received the energy efficiency tag A are sold for 4–12% more than similar houses with energy tag D.

The French Eco-Districts and Their Social Aspects

About 70% of all greenhouse gas (GHG) emissions originate in urban areas. While national policies strongly affect local energy consumption, as well as the generation and disposal of pollutants, city planning and policies can help reduce GHGs, promoting sustainable development.

In 2008, the European Commission launched the *Covenant of Mayors*. Signatories agreed to exceed the EU target of a 20% reduction in GHG emissions (from 1990 levels) by 2020. With more than 5000 signatories, the Covenant is the leading movement in Europe supporting local and regional authorities to achieve this goal.

Some European cities started focusing on climate change well before signing the Covenant and made significant inroads in GHG reduction: the successes of Stockholm, Malmö, Freiburg, Copenhagen, and others have been widely publicized. One practice common to these cities is establishing eco-districts—neighborhood scale developments addressing climate mitigation as well as energy savings, and the management of waste, transport and water resources. The establishment of such eco-districts may indirectly raise development standards throughout a city, by setting a benchmark to which others aspire. There is a growing literature on eco-districts, but little systematic analysis of their role in attracting affluent populations.

In French eco-districts, or *eco-quartiers*, social diversity is a mandatory component. According to eco-quartier framework requirements (2009), there should be a socio-economic, generational and cultural population mix within the neighborhood – in stark contrast to the large housing projects built for the poor from 1956 on (mostly in the 1960s and 70s) to deal with the serious housing crisis that France faced after the Second World War (Blanc 2004). The construction of three million dwelling units within 10 years was expected to solve the crisis.

In the 1960s, these French suburban social housing estates accommodated their first residents. They are multi-ethnic, hosting poor whites, blacks and Arabs (mainly from French ex-colonies) who are severely stigmatized. Since the late 1970s, French middle-class families have been leaving them, usually to acquire single-family houses in more distant suburbs. Those who remain, mainly the second generation of blacks and Arabs from French ex-colonies, are generally those who cannot afford moving out. These neighborhoods are characterized by a high rate of poverty and unemployment.

Whereas discussions about segregation by social class or ethnic group have a long history in America, France has tried to deny its own segregation practices, as they expose its failure to implement the French republican ideals of Equality and Fraternity. The spectacular riots in the Fall of 2005 dramatized ethnic segregation, stunning public opinion. The French republican model of integration had been brutally shaken by an obstinate reality (Dikeç 2007).

In response to the growing spatial concentration of poor and minority ethnic groups in the 1980s, social mix came to be viewed as a means to regenerate social housing estates in Western Europe. In France the national urban renewal programme (Programme National de Rénovation Urbaine), launched in 2003, stated the aim of "encouraging social mix and sustainable development". As in other countries, the demolition of French social housing in deprived neighborhoods aims to disperse poor populations while new, more expensive private housing developments are built to bring in middle-class households (Lévy-Vroelant 2007). This policy assumes that more social diversity in poor neighborhoods will improve their economic viability, as well as the reputation and livability of the area. Furthermore, the policy assumes that such diversity generates social interaction and opportunities benefitting the poorest populations (Tunstall and Fenton 2006).

These policies, however, have sparked widespread scholarly debate. The intensity of negative "neighborhood effects" due to poverty concentration is disputed. Moreover, some scholars doubt that social mix strategies benefit lower-income groups. Studies of social interaction in regenerated neighborhoods, reveal a loosening of social bonds and increased group conflict.

These assertions challenge the positive connotation of social mixing, which according to the Habitat and Housing dictionary, aims "to allow class coexistence within the same urban unit" (p.297). It was defined as a "miracle cure" (Baudin 2001) for contemporary cities, and was seen to be "a privileged means of rebuilding social ties or cohesion" (Bacqué, p.298). Diversity would thus "make society" (Donzelot et al. 2003), locally and nationally.

Despite the buzzword of 'social mix' being so frequent in urban policy statements and especially in those of the "eco- neighborhoods", it is viewed by some as "closer to a myth than to a scientific logic". Kirszbaum reminds us that this buzzword is an expression of "nostalgia for the golden age, the pre-Haussmanian Parisian building, in which each social class occupied its own floor, but was living in peace and harmony with the others" (Kirszbaum 2008, p.45). This nostalgia could be defined as a "republican mythology of democratic equality", but also as a "strange historical amnesia" (Simon 1995), ignoring chronic class enmities. Social diversity is presented as a solution to a wide range of issues, affecting both republican equality and local social cohesion, and seen as "a simple answer to at least two complex questions": that of the assumed social effects of cohabitation (social links, integration), and that of the public authorities' ability to affect it (Lelévrier 2006, p.5).

Researchers have shown that improving social cohesion through social mix is perceived as a means of strengthening social control, reducing deviant behavior and personal insecurity (Kearns and Mason 2007; Sampson et al. 1997). The implementation of a social mix policy assumes that the middle classes will play an integrating role for the working classes and that the presence of the former in these neighborhoods will generate "social capital" (Flint and Kearns, 2006). However, some observers believe that decision-maker optimism about the effects of social mix is unjustified (Lelévrier 2006), and that implementing it through housing diversification programs will not improve social cohesion, nor local social relations (Bond et al. 2012). Moreover, group conflicts can be exacerbated by differences in values and norms (Authier 2007).

Social mix in the eco-quartiers is meant to be ensured through housing diversity, i.e. in terms of price, dwelling size, typology (individual vs. collective), social/private housing or modality of access (purchase or rental). The following case studies illustrate how different municipalities have distributed their eco-quartier housing to attract the middle class under the guise of social mix.

Case Studies

By 2017, 26 French municipalities had received the national eco-quartier label. We have listed localities according to their socio-economic status (see the table in appendix – villages and overseas cities are not included[3]).

The three case studies in France. (Map: Machline 2017)

France is one of the more affluent countries in the OECD, but exhibits significant poverty and profound inequalities of income, wealth, and life outcomes. The past several years of rising unemployment have worsened these circumstances. This variation often occurs across lines of ethnicity and national origins; immigrant communities in France tend to have significantly higher levels of poverty, unemployment, and health disparities, often concentrated in specific suburbs of major cities (Lautre 2018).

Three eco-quartier case studies were selected for closer examination, each representing a different category of municipality: (1) The Parisian metropolis with a high average income (though a few poor neighborhoods remain, mostly in eastern Paris, in the 18th, 19th and 20th districts); (2) the remote Parisian suburb of Bretigny Sur

---

[3] The villages: Forcalquier, Hede-Bazouges, La Chapelle Sur Erdre, La Riviere, Sainte-Croix-aux-Mines, Longvic, Morez, Change. In the French overseas department of La Reunion: Saint Pierre.

Orge, with a high share of low-income social housing; and (3) the municipality of Reims, with a high rate of poverty and large income disparities between poor and rich neighborhoods.

The logic of case study selection was to allow for an analysis of how municipality location (core-periphery), and socio-economic status, affect the provision of 'green' and affordable housing, within the context of local housing policy and how its green building concept was developed. (Due to budget constraints we only considered the northern part of France, within a radius of 300 km from Paris).

Data collection included interviews and participant observation in key professional events, as well as analysis of documents from a variety of sources (policy-making bodies, NGOs, interest groups, experts, the media).

The Major City: Paris

The map in Fig. 1 represents eco-quartier distribution among six of the twenty administrative districts (*arrondissements*) of Paris: Gare de Rungis (13th), Clichy-Batignolles (17th), Pajol (18th), Frequel-Fontarabie (20th), Boucicaut (15th) and Claude Bernard (19th). In total, they so far include only 0.6% of the Paris population

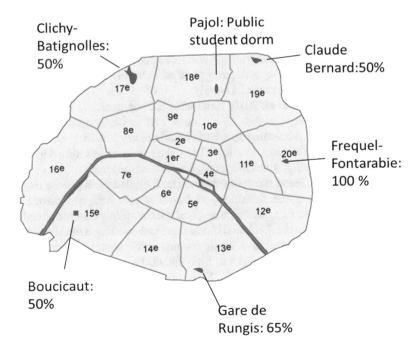

Fig. 1 The location of eco-quartiers in Paris, and the percentage of social housing in each. (Source: Machline et al. (2016))

Paris Municipality (2015): Personal communication and half of their total housing units are social housing. The Claude Bernard eco-quartier includes 50% social housing, 25% private rental housing (controlled rent) and 25% private dwellings of various sizes to ensure social diversity (as well as a home for the elderly, for generational diversity). It was stipulated in the eco-quartier label file that social housing for the upper middle-class should compensate for the high share of low-income dwellings in the neighborhood (see the socio-economic classification in France, Annex Table 2). In 2009 the private apartments were sold for €6250/sq.m. (according to the developer's representative), well above the 19th arrondissement average of €4900/sq.m., but slightly below the new apartment average price at the time (€ 6500/sq.m.). According to the neighborhood developer representative (personal communication, 2016), the promoter had to reduce apartment prices to sell them, because "no one would have bought an apartment in that neighborhood – next to the highway, without public transportation, and with so many prostitutes and drug dealers".

The Eco-Quartiers: Affordable Housing or Hidden 'Green Gentrification'?

Figure 2 shows that the share of social housing units for low income families is minimal in all eco-quartiers: 5% in Claude Bernard, 10% in Clichy- Batignolles, 3% in Gare de Rungis and 5% in Frequel Fontarabie. According to Machline et al. (2016), all eco-quartiers are located in statistical block areas already including a substantial share of social housing units – and three out of the six have 90% or more such units. Moreover, no eco-quartier was planned in blocks lacking social housing (Machline et al. 2016), to increase its supply.

According to Machline et al. 2016, four out of six eco-quartiers were planned in the lowest socio-economic status block areas and arrondissements[4] (Category 3–13th and 18th–20th). According to a Paris Municipality Planning Unit representative (personal communication, 13 April 2016), the eco-quartiers aimed to increase the share of private dwellings and attract higher income residents.

As mentioned, the Claude Bernard eco-quartier included 50% private dwellings and 25% middle-class social housing – and thus it was designed primarily for middle-class residents. In fact, the neighborhood's share of low-income social housing is only 5%, whereas in the statistical block areas surrounding it the share exceeds 90%. The 19th arrondissement as a whole includes 35% social housing while the average annual household income is less than €30,000 – very low compared to the Parisian average (Machline et al. 2016). Thus, at least 75% of the housing units in

---

[4]Category1: Census blocks with a high median income and high share of housing ≥100 m², self-employed, craftsmen, managers and people with high educational degrees; Category 2: Census blocks with a high share of managers and academics; Category 3: Census blocks with a high share of non-graduates, blue collar workers, employees, subsidized housing, single parent families, unemployed, people with basic or intermediate qualifications, non-owners, and foreigners.

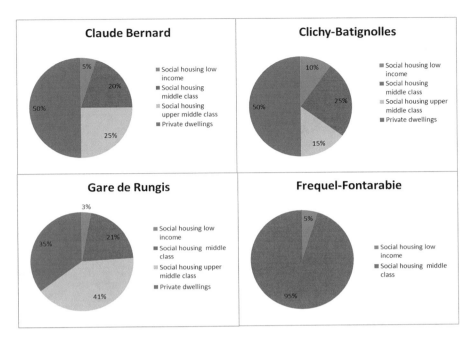

Fig. 2 Share of social housing by category in selected Paris eco-quartiers. (Source: Machline et al. (2016))

the eco-quartier will be unaffordable to the local (19th arrondissement) population. According to the social housing company representative (2015, "personal communication") "the shares of social housing categories were decided upon by the municipality" (Fig. 3).

Dwellings in Claude Bernard, public or private, fail to provide 'affordable housing prices' (as stipulated in the certification file) to pre-existing residents of the 19th arrondissement. According to the Paris Municipality Planning Unit and housing authority (2015), the aim was to attract better off residents: "The aim is to balance the socio-economic level of the population" (Head of the Housing Authority of Paris 2015: Personal communication). According to a representative of the developer (2015), veteran residents were strongly encouraged to move out.

The private dwellings in Claude Bernard were sold at 4% less than the average 19th arrondissement prices for new housing. Moreover, according to the manager of the planning company (2015): "At the beginning it was not easy to sell the apartments. The first home buyers were very brave to move into a neighborhood where they were exposed to prostitution and drugs on a daily basis!"

The situation is similar for the Gare de Rungis eco-quartier (13th arrondissement), where the share of social housing is 65% but the upper middle-class public housing units constitute 63%, while the low-income units are only 4.4% of the social housing stock (Fig. 2). In the Frequel-Fontarabie eco-quartier, which included 27 of the 1030 Parisian substandard buildings (Observatory for Building Degradation

Offices building   **Elder house and private dwellings**
Pictures: Machline, 2014

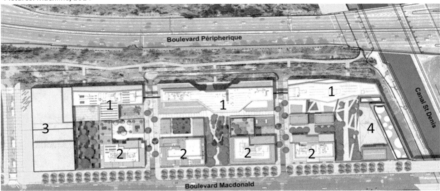

1- Offices (41 000m2); 2-Housing (27 000 m2), Elder house (6200 m2), Commerce (7000 m2); 3- leisure
(10 000m2); 4- Public equipment

Fig. 3 The Claude Bernard eco-quartier. (Source: Machline et al. (2016))

Prevention in Paris 2012), 95% of the social housing is meant for the middle-class
(Fig. 2). However, before the establishment of the eco-quartier, all of the housing
stock was for low-income households (Paris Housing Authority 2014: Personal
communication). Thus, 38% of the social housing built in Paris since 2001 was
intended for the upper-middle-class, which did not match the demand: among the
100,000 applicants for social housing in Paris, 75% earned below the low-income
housing limit, while only 4% could apply for upper middle-class housing (APUR
2007). In recent years low income public dwellings have represented only 15% of
the social housing built in Paris (Clerval 2010). At the city scale, between 2001 and
2013, the newly built social housing consisted of 25% for low income and 37% for
middle income, while the share for the upper-middle-class was even slightly larger,
at 38%. This latter city scale ratio is substantially lower than that of the eco-quartiers
Gare de Rungis (63%) and Claude Bernard (50%) – and since such social housing
promotes gentrification in the poorest districts of Paris, the eco-quartiers foster
'eco-gentrification', at the expense of the poor. While working-class households

face growing difficulties in finding public or private dwellings in Paris, public policies have improved housing conditions for those who can afford to stay in the city or move into it (Clerval and Fleury 2012).

A Remote Suburb: Bretigny Sur Orge

The Clause Bois Badeau eco-quartier is located in the small municipality of Brétigny sur Orge, 35 kilometers south of Paris, in the department of Essonne, in the Ile De France region,[5] on line C of the RER.

In the second half of the twentieth century, Brétigny-sur-Orge, like most municipalities in the Paris region, experienced rapid population growth, going from 3673 residents in 1946 to more than 12,000 in 1968 and 21,650 in 1999. Later censuses show that in the next decade the population stabilized with 21,837 residents in 2007. The territory of Brétigny-sur-Orge remains mostly farmland, cultivated by the Clause farming company. In 2000 eight farms remained active, while in 2006 136 people still worked in agriculture. Unemployment was low (8.4%) in 2006, but 17% of employees worked on precarious contracts. Employees and workers are widely represented in the commune, while managers, craftsmen and traders are fewer than in the rest of the department. However, there are disparities, as 55% of the tax households paid solidarity tax on wealth,[6] while only 67.9% of the households paid income tax (the average net tax income in the municipality being about 25,000 euros per household). In 2006, only 51.6% of the Brétignolais owned their dwellings, while 25.7% lived in HLM's. In 2011, the median tax revenue per household was € 360,000, and most (76.8%) of the population worked in another municipality. Bretigny Sur Orge is a middle-income municipality (the national median income being €20,369- INSEE, 2014) (Fig. 4).

The eco-quartier will include 2400 housing units, with 30% of them defined social housing. The total population in Bretigny is 23,000 habitants, and thanks to the eco-quartier it is projected to reach 30,000 within 15 years. In 2016, 700 housing units were built, 46% of them social housing (for the first phase). Overall, the municipality of Brétigny sur Orge has 32% social housing units. The eco-district was built on 42 hectares, mostly agricultural fields, and project development began in 2006. So far, the 46% share of social housing includes 8% for low-income households and 29% for middle-class households (Fig. 5).

---

[5] The French Revolution abolished the historical French provinces replacing them with an arbitrarily drawn system of departments (100), each responsible to the French government, thus increasing centralization. Worried that power had become too centralized in Paris, the law of decentralization (1982) established a system of 22 regions with boundaries similar to those of the provinces of old, to promote local cultural differences and economic development. In 2016 the number of regions was reduced to 13.

[6] The solidarity tax on wealth is an annual direct wealth tax on residents with assets in excess of €1,300,000, (since 2011).

Fig. 4 The Clause Bois Badeau eco-quartier. (Source: Machline (2016))

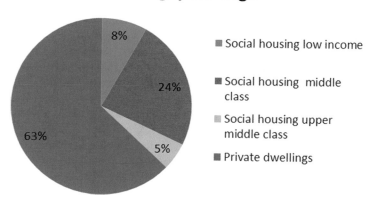

Fig. 5 Housing distribution in the Clause Bois Badeau eco-quartier. (Source: Municipality of Bretigny Sur Orge (2016): Personal communication)

According to the development company's project manager, the reason for building 46% social housing – rather than the originally planned 30% – was the 2008 real estate crisis. Some promoters had to cancel their participation, and thus, to implement the project, the socialist-led municipality decided to plan more social dwellings, as social housing companies are funded by the State. The project, which was to start with private apartments, began with social housing units. However, according to the Head of the Planning Unit of Bretigny sur Orge, the new neighborhood's start with social housing gave it a bad image in the town.

Bretigny is mainly composed of low-density neighborhoods; the population was scared that Clause Bois Badeau would become a "dangerous vertical slum", like others in the Essonne department, for example in Corbeil-Essonnes, Evry and Grigny. There was a feeling that the entire municipal budget would be spent on this new neighborhood. The loss of the main green and open spaces of Bretigny was also hard to accept. According to the Development Company Manager: "The owner of the private rental buildings encountered difficulties in finding renters – for instance, the cheapest four room unit (100 sq.m. on average) costs 2400 euro/month and should not be above 1800 to be affordable for a family." Apartments for sale in the new neighborhood (about 3500 euro/sq.m.) are cheaper than those around it (3800 euro/sq.m.). Thus, the building of a 'green' neighborhood in Bretigny Sur Orge had no gentrifying effect. The newly elected right-wing municipal team (in 2014) canceled a social housing operation before its implementation and, since the election, only private dwellings have been designed. The aim is to attract a more affluent population and even plan single-family house projects (Development company project manager, personal communication 21/04/2016).

According to the development company's project manager, most home buyers are young families from Bretigny or surrounding cities, purchasing their first apartment. The municipality wanted at least 60% of the homebuyers to be residents (to limit the share of outside investors). Thus, the real estate companies selling the apartments were asked to prevent that share from rising above 40% (Head of the Urbanism Unit, Municipality of Bretigny, personal communication 18/04/2016).

Within the eco-quartier, a central urban heating network was installed, consuming wood pellets (considered cleaner than other heating fuels). However, according to the Head of the Urbanism Unit, the residents complain of thermal discomfort and construction defects. Most buildings fail to attain the planned energy efficiency, and even if overall consumption is lower than for conventional buildings, the residents' monthly bills are not lower, as the maintenance cost of the 'greener' heating system is higher than that of conventional natural gas heating. To avoid construction defects, the municipality has tried to ensure that the promoters improve their inspection skills and the Bretigny Mayor makes it a point to be present for each building inauguration and warn the promoters that "if there are construction defects, they will hear from him" (Head of the Urbanism unit, municipality of Bretigny, personal communication 18/04/2016).

A Remote Periphery: Reims

Reims, in the Marne department of the Grand Est region of France (previously in the Champagne-Ardenne region), lies 129 km east-northeast of Paris. The 2013 census recorded 182,592 dwellers in the city proper and 317,611 inhabitants in its metropolitan area. Historically, the Notre Dame de Reims cathedral has been the place of coronation of French kings. The bubbling Champagne, produced in the Reims region since the end of the seventeenth century, has become one of its economic assets. The city of Reims has undergone rapid economic growth between 1960 and 1975, thanks to new industries (some of them owned by IKEA). The period was marked by notable spatial expansion and a 54% demographic growth rate (from 131,000 in 1954 to 200,000 in 1975). Like most French cities, Reims subsequently experienced economic and demographic stagnation.

The arrival of the TGV high-speed train in 2007 provided a new boost to growth, putting Reims at just 45 minutes from Paris, 35 minutes from Marne-la-Vallée[7] and 30 minutes from the Roissy-Charles de Gaulle International Airport. Reims would like to host Parisian companies wishing to relocate to lower operating costs. A new tramway line was built in 2011, running through the Reims city center and serving its two most important neighborhoods, the Orgeval district in the north and the Red Cross district in the south. Notwithstanding these growth enhancing steps, Reims still belongs to the poorest third of French municipalities.

The construction of large housing projects began in the 1960s. One of the most important HLM neighborhoods is the Croix Rouge district (divided into three sectors: Nord, Pays de France and Croix du Sud), launched in 1967. Located in the south-west of the city, this district had 25,000 inhabitants by 2010, almost as many as the entire city in 1962. The faculties of Law and Letters were established in the north of the Croix Rouge District. The Faculties of Medicine and Pharmacy are also located in the district, at the edge of the University Hospital Center. These last two public facilities are in the Pays de France sector, while the Faculties of Law and Letters are in the Northern sector (University sector).

The Pays de France Eco-Quartier: Refurbishing a Low-Income Social Housing Neighborhood

The Pays de France eco-quartier is located in the Croix Rouge district (with a population of 20,000) and has about 4000 residents. It is part of a larger urban renewal plan drawn in 2004 and implemented from 2009 to 2015. The project was funded by the ANRU, the European fund FEDER, "Caisse des Dépôts et Consignations", the Champagne regional council, the Marne department, the Reims municipality and the "Foyer Remois" social housing company.

Historically the "Pays de France" district was a social housing neighborhood, built like many others in 1960–1970. The housing company is FOYER REMOIS. The

---

[7] Marne-la-Vallée is one of the five "new towns" established by General de Gaulle in the Paris region at the end of the 1960s to host new populations and improve living conditions. Public transportation and highways were provided to allow commuting to Paris. Today, Marne-la-Vallée hosts multinational corporations as well as research centers and a dense network of small and medium-sized businesses. It represents an important employment source.

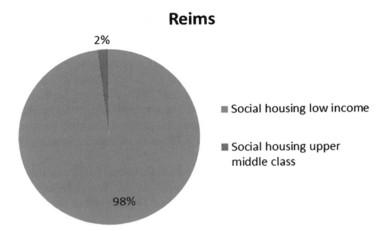

Fig. 6 Social housing distribution in the Pays de France neighborhood. (Source: Municipality of Reims (2016): Personal communication)

Pays de France neighborhood is the most fragile area of the Croix Rouge district, with an impoverished population and a severe lack of local businesses. A study by the municipality in 2006 lists the following perceived dysfunctions:

- The urban design is inappropriate, even obsolete (slab planning due to functionalist conception), with a separation between pedestrian and car traffic;
- Shops and services are lacking;
- Public and private open spaces are in bad shape;
- There is no functional diversity (almost only housing);
- The population is fragile, with higher rates of families in economic difficulty and of young people below 25 years of age, than other neighborhoods in Reims[8];
- A feeling of insecurity (Fig. 6)

The neighborhood only included low-income social housing units and has been refurbished with ANRU[9] subsidies. Among the 530 apartments, 158 were torn down and some residents were relocated in other social housing neighborhoods in Reims. According to the certification file, to promote social mix, 45 units for upper middle class and 23 social dwellings for private ownership were to be included. However, according to the Head of Reims City and Housing Policy Department and the Head of Social and Urban Development Department of the Foyer Remois social housing company (19.08.2016, personal communication), only 17 upper middle-class housing units were built and remained empty, as their rent is even higher than in the private sector. Furthermore, the image of the neighborhood detracted from the

---

[8] The unemployment rate for French youth is 28.3%, compared to 8.7% for the total population.

[9] The National Agency for Urban Renewal (ANRU) monitors the national urban renewal program (PNRU). Objective: To renovate 530 neighbourhoods by 2013, totaling 40 billion euros of investment.

apartments' attractiveness. "We made upper middle-class apartments, but it was no success. Still, we believe that the neighborhood's great location will attract new populations. We are optimistic. At present, there is no population renewal. Until we renovate the entire district (Croix Rouge), its bad reputation will linger on. It will probably take an entire generation to change the image. Even if the eco-quartier buildings are high standard, they are surrounded by very low-quality buildings. The social mix aspect was one of the weaknesses of our application. We could not satisfy all requirements, given that we first planned the projects and only then applied for the label." (2016, Head of Reims City and Housing Policy Department, "personal communication").

All buildings received the BBC certification. However, to pay for the refurbishment, the social housing company raised rents by about 50 euro/month. According to the Head of the Social and Urban Development Department of the Foyer Remois social housing company, "thanks to the BBC certification requirements, the final energy consumption decreased by 30–50%. Thus, even if the rent increased, the residents pay little more than before, 5 euros more on average). According to the Head of Reims City and Housing Policy Department, when the rent increase is too high (as it has been for a few families in Pays de France), the municipality has subsidized it during the first 4 years (20 Euros/month during the first year, then dropping to 15, 10 and 5). In Reims, the eco-district had no gentrifying effect and brought 'green' housing to the poor (Fig. 7).

Conclusion

The specificity of the French eco-districts remains the social mix requirement, countering socio-spatial segregation. Avoiding the repetition of past 'mistakes' (such as concentrating poor people in large HLM projects) is essential in the view of French planning authorities. The eco-quartiers are usually located in substandard districts of large cities or average income municipalities, and their role is to attract middle class people, seen as guarantors of social cohesion. However, social mix goals are barely reached. In affluent cities where property prices are high (like Paris, its suburbs and some large cities), the municipalities build eco-quartiers in substandard neighborhoods to attract middle class families. In average cities (such as Bretigny Sur Orge), some municipalities have implemented more social housing than planned, to obtain State subsidies and loans for developers (thus the greening was motivated by budget considerations) – but can still privilege the middle class in that housing's allocation. The high share of social housing is due to these municipalities' difficulties in attracting middle-class customers to buy the private apartments and to their willingness to accept some poor residents, if that's what it takes to encourage further construction. In the poorest French towns, like Reims, the eco-quartiers will improve living conditions in substandard neighborhoods but remain irrelevant to social mix.

Fig. 7 The Pays de France Eco-quartier. (Photo: Machline (2016))

Attracting well-to-do populations is only possible in Paris, including some of its closer suburbs and other affluent cities (for example Bordeaux), where the share of low-income populations in the eco-quartiers will be minimal. In the wealthiest large cities, 'green' housing has a gentrifying effect. However, in the municipalities of middle and lower-income, eco-districts cause no gentrification.

Discussion-Conclusion

This chapter discusses key findings and concludes the study, which considered the socio-economic impact of 'green' building in Israel and France and examined social aspects of sustainable urbanism in the two countries.

Policies for Promoting Affordable Green Housing: France vs. Israel

France subsidizes public housing construction through national and municipal funding. The SRU law (see chapter "Green Building in Europe"), launched in 2000, confirms the right of every citizen to decent housing. Every municipality numbering more than 3500 residents must ensure 20% of social housing at the very least. As explained in chapter "The French Case Studies", social housing is also available in Denmark and Sweden, where it is often municipally owned.

As an EU member, France has access to funds earmarked to help implement the European Energy Performance of Buildings Directive (EPBD). A low consumption building label (BBC-see chapter "Green Building in Europe") has been established in 2005. Today, half the certified apartments are public housing. Meanwhile, one way to implement the EPBD has been establishing a national framework for eco-districts, whereby municipalities receive budgets from the EU (through the State) to subsidize 'green' apartment construction.

While other EU countries promote social as well as green housing, France is unique in the way it integrates both policies, unlike the Scandinavian countries where green building is no part of the social housing policy. The French eco-quartier guidelines offer a potential for economic development and social and functional diversity.

© The Author(s), under exclusive license to Springer Nature Switzerland AG 2020
E. Machline et al., *Green Neighbourhoods and Eco-gentrification*,
SpringerBriefs in Environmental Science,
https://doi.org/10.1007/978-3-030-38036-6_6

In Israel, the allowed renters of publicly funded apartments to purchase them at reduced prices. Meanwhile the social housing stock, which was already restricted beforehand, was not renewed not even from the funds collected through resident apartment purchases. Moreover, in the 2000s, the government sharply reduced social expenses, shrinking Israel's welfare state, and no further social housing has been built since then (2019).

A standard to certify buildings with "reduced environmental impact" – IS 5281 – was established in 2005 but adopted as a voluntary measure and only marginally implemented. Since 2008, 18 of the largest cities in Israel have joined the International Council for Local Environmental Initiatives' (ICLEI) Cities for Climate Protection Program (CCP), and signed the Forum 15 Convention, committing them to reduce their Greenhouse Gas (GHG) emissions to 20% below the levels in year 2000. In June 2013, those cities decided to gradually turn the green building standard into a mandatory one. However, a neighborhood-scale 'green' certification has been developed very recently and is not yet implemented. Moreover, in the Israeli 'green' neighborhoods there is no policy mandate for affordable housing. The TA 3700 project in Tel Aviv is to be the first attempt to apply sustainable neighborhood design principles (following LEED-ND) in master plan elaboration. In Neve Sharett and Yavneh, the 'green' neighborhood label is not due to the urban plan but to the presence of SI 5281 certified buildings. In the Dimona neighborhood there is no environmental certification, even though it is marketed as "in keeping with Israeli Green Building Standards".

'Green' Neighborhoods and Affordability

Our study shows that in locations where housing prices are high, 'green' certification makes real estate even less affordable and even countries with a strong social policy can only mitigate that trend.

In Israel we found that 'green' certification raises apartment sale prices by 3–14% (about 7% on average), with higher values in the northern periphery than in the center (i.e. the Tel Aviv metropolitan area).

Strictly parallel data representing the 'green value'in France were unavailable, but one indication of its magnitude is a statistical comparison showing that apartment prices in buildings labeled with energy tag A[1] (specifying annual energy consumption of less than 50 kWh_{pe}/m^2) were higher by 4–12% than those ranked with tag D (with annual energy consumption of 151–230 kWh_{pe}/m^2). In France the 'green' value is also clearly larger in the periphery than in Paris (see map in chapter "The Socio-Economic Impacts of 'Green' Building in Israel: Green Building as an Urban Branding Tool"). In sharp distinction from Israel, however, more than half of

---

[1]According to the French law, homeowners must establish an energy performance diagnostic (graded from A to G) to be able to sell their apartment.

the environmentally certified apartments in France are marketed as public social housing – meaning that ostensibly, 'green' apartments are available to relatively low-income residents, as well as to those who can more easily afford a 'green premium'.

As the law fails to specify the shares of social housing types, units for low-income families are very few in attractive locations, like Paris (5% average). More generally, in affluent cities (like in Paris, its well-to-do suburbs and some large cities), municipalities build eco-quartiers in substandard neighborhoods, to attract middle-class families. In average cities, like Bretigny sur Orge, municipalities tend to implement more social housing than planned. This is not due to a wish to host low-income populations in the eco-quartiers, but rather to the difficulty of attracting middle-class customers into buying private apartments on the one hand and on the other hand, to the ability of social housing companies to obtain State subsidies and loans, while privileging apartments for the middle-class. In the poorest French towns, like Reims, eco-quartiers will improve living conditions but remain irrelevant to social mix. Attracting well-to-do residents is only possible in Paris, its closer suburbs and other affluent cities, where low-income families are few in eco-quartiers. Indeed, in the last revision of the Eco-quartier label guidelines (2016),[2] the ministry deleted the requirement that the local authority verify "whether housing location and prices are appropriate to household's income", considered as "too complicated to fulfill for municipalities".

In Israeli 'green' neighborhoods there is no policy mandate for affordable housing. To the contrary, the poor are kept out or encouraged to leave. In Tel Aviv, apartment prices and rents have more than doubled in Neve Sharett, following the 'greening' of the neighborhood. Thus, no renters (40% of the residents) could afford to stay in it. Moreover, 75% of the owners declared that they would not live in their new 'green' apartment, as its operation costs are too expensive (higher energy and water bills, municipality taxes and maintenance charges), even though it is stated that the 'green' building certification would reduce energy consumption by 30%. In other words, energy efficiency targets have not been reached.

In Yavneh and Dimona, "green" apartments are more expensive than other new apartments and the municipalities acknowledge that they are meant for middle-class residents. In Yavneh, both housing prices and socio-economic data show that the 'green' neighborhood indeed serves middle and upper middle-class residents, as intended by the municipality. Even in the plan for a LEED ND neighborhood in Tel Aviv, households below the 7th socio-economic cluster are not eligible for subsidized housing – which is meant for the middle-class, not the poor.

Building according to the SI 5281 standard was possible in Yavneh, but not in Dimona. It seems that in peripheral locations, where real estate prices are low, builders are unwilling to apply the standard. The new national plan for affordable housing (*Mehir La'Mishtaken*, which offered apartments for about 20% below

---

[2] Le referential Eco-quartier, 2016, Ministry of Housing, the Equality of territories and Rurality.

market prices) was to be implemented in both 'green' neighborhoods, but only in Yavneh were the subsidized units to be built according to the SI 5281 standard.

In both Tel Aviv and Yavneh, new plans have been approved for LEED ND neighborhoods. However, it is doubtful that LEED ND certification will promote green and affordable housing in Israel. As pointed out by Szibbo (2016), in 60% of the LEED ND projects in the US, there is no affordable housing. In the LEED ND guidelines, in the 'Neighborhood Pattern and Design' category, only seven optional points (out of a minimum of 40 points for the lowest certification level) are allocated for affordable housing, under the sub-category: 'Mixed Income Diverse Communities'.

Given that Israeli housing prices are considered unaffordable by at least 70% of households (see Introduction) and that 'green' certification (according to SI 5281) significantly raises prices (see chapter "The French Case Studies"), the implementation of the LEED ND framework might exacerbate an acute national problem. A truly sustainable urban neighborhood requires more than energy-efficient design; it also must be economically viable and socially acceptable to residents.

In both Paris and Tel Aviv, 'green' neighborhoods have been built in poorer neighborhoods where housing prices had been affordable. The cities' attractiveness allows 'green' building to promote gentrification– attracting wealthier people to poorer neighborhoods. However, in Reims the eco-quartier could only help improve the housing conditions of the current low-income population, while in Dimona, low housing prices prevented the apartments from being built to green (SI5281) requirements – and building companies encountered difficulties in marketing them, as only local families purchased housing units or lots.

Bretigny sur Orge is intermediate between 'core' and 'periphery'. Its moderate distance from Paris (50 km) makes it more attractive than Reims or Dimona. However, it is not close enough to Paris for gentrification. Thus, private apartments were hard to sell, while the share of public social housing turned out higher than expected, due to a lack of buyer interest, even though apartments are cheaper than in surrounding neighborhoods.

In sum, 'green' construction may easily lead to gentrification – raising property values and rental costs, stimulating new construction or renovation, upgrading the housing stock, causing local population turnover, and bringing in higher status residents. However, the gentrification process fundamentally differs between the two countries. 'Green' gentrification appears to be occurring in France, despite a declared intention to prevent it – while in Israel, 'green' gentrification appears to be the underlying intent. In terms of similarities, it may be seen that in both countries 'green' gentrification (1) occurs in wealthy but heterogeneous cities, and (2) does not appear in poor localities. However, only in Israel does 'green' gentrification occur in medium-income cities.

'Green' Neighborhoods or 'Greenwash'?

Our study has shown that in both France and Israel, the environmental benefits attributed to 'green' buildings, simply because they have been certified as 'green', are often questionable. For example, certified green buildings in Israel are purported to achieve "savings of approximately 30% in energy use" (According to the Israeli Ministry of Environment-see Appendices), but the mandatory standard SI 1045 already ensures that the requirements of the least demanding (one star) level of the 'green' certification are met in terms of energy (Goulden et al. 2015). In other words, the SI 5281 buildings are not necessarily more energy efficient than conventional buildings, assuming that building codes are respected. As shown (see chapter "The French Case Studies"), only a small portion of the green neighborhood residents surveyed report significant energy savings, which in any case hardly appears to be a prime motivation of 'green' apartment buyers.

In France, while energy targets were not met in the assessed eco-quartiers, improvement was achieved in comparison to the high-energy demand of apartments built before the 2005 regulations. In Paris, for example, a Ministry study found annual consumption levels of 80–100 kWhep/sq.m., nearly twice as high as the promised targets, but less than a third of the consumption in social housing built in the 60s–70s.

Beyond its technical and energy efficiency indicators, an urban development project is also said to be sustainable when it generates "a harmonious living environment, reduces social inequality, and improves quality of life". According to debates about sustainability are no longer limited to environmental aspects, but also incorporate economic and social dimensions. Socially sustainable projects should provide social infrastructures and job opportunities, and ensure easy access. However, in Israel such socio-economic aspects of urban sustainability have hardly been tackled.

The marketing of 'green' building projects, residential complexes, and neighborhoods has become ubiquitous in recent years, suggesting that environmental buzzwords have become effective marketing tools. It is often difficult, however, to distinguish between the actual environmental value and the 'green' image being marketed. One motive behind the development of the French eco-quartier label, was to prevent the marketing of a non-green real estate project as an eco-district – i.e. to avoid 'greenwashing'.

Conclusion

An underlying premise of this study is that gentrification under the banner of green building runs counter to the principles of sustainable development, which rests on social and economic, as well as environmental, pillars. 'Green' and affordable housing implementation are, in both countries, directly connected to affordable housing policy: France still subsidizes public rental housing, while Israel has not done so for decades. The comparison of approaches adopted by the two countries, albeit within different contexts, illuminates the possibilities of providing 'green neighborhoods' to a moderately large cross-section of the population. The French State has made this a declared priority, reminding one of the socialistic ideals that once guided development in Israel. But declarations are hardly enough to bring meaningful change, and the French eco-district is merely a small step towards turning the goal into a reality. In both countries, 'green' building is mainly practiced in middle-class urban areas – and in low-income areas of wealthy cities, to attract middle-class residents. Neither country has built 'green' housing in wealthy neighborhoods and it is only in France that we find 'green' housing in poor localities – though even there, the ideal of a green and socially mixed neighborhood remains unrealized.

The comparison of the two countries approaches' has shown that it might be possible to provide 'green neighborhoods' to a sizable cross-section of the population. Even in Nordic cities internationally known for social cohesion and relative social equality, polarization and fragmentation are seen as severe challenges. The capital cities of Sweden and Denmark show socio-economic segregation in their eco-districts.

The socio-economic consequences of 'green' building are similar in France and Israel for wealthy municipalities. However, for average and poor municipalities, the situation is different: In France, the authorities fund a large share of social housing (due to the lack of private investors) and thus there is no gentrification. In Israel, in middle income localities where there is a potential for gentrification, we do find 'green' building. In poor localities there is none.

The sale price of an apartment in a certified green building is significantly higher than justified by either additional construction costs, or energy and water saving potential.

The French policy promotes social diversity and the construction of 'green' public social housing in the eco-districts. Thus, there is an ostensible effort to build housing that is both 'green' and affordable. However, that policy is applicable only in locations where housing prices are affordable. Thus, in affluent and average municipalities, the share of 'green' social public housing actually available to low-income groups is very small – since most public social housing is ultimately allocated to higher-income groups. Such is the case in eco-districts, which usually include about 40% subsidized units in their housing mix, but mainly supply them to middle-class residents. Thus, in attractive French localities, policies of 'green' housing for the poor are being subverted and turned into 'greentrification' of lower-class neighborhoods.

Regarding the potential application of French policy tools in Israel: First of all, Israel's government is unlikely to re-engage in building subsidized social housing ('green' or otherwise) in the near future. In 2016, Israel's public social expenditure as a proportion of its Gross Domestic Product was 16%, half the 32% invested by the French government. Israel recorded a government debt equivalent to 62% of the country's GDP in 2016, while in France the equivalent figure was 96%. According to *Ha'aretz*, 22/10/2017, (https://www.haaretz.co.il/1.4530532-), Israel finished

2017 with a budget surplus of 10 billion NIS (2.8 billion $US) but has incurred a deficit in 2018-1019. That budget funded defense expenditures and lower taxes for higher income citizens. In 2018-2019 a serious deficit may cause a further decline of the social budget.

Another context difference is that unlike France, Israel cannot benefit from European Union subsidies for 'green' housing. In addition, the French state is highly centralized, and traditionally plays a major role in implementing such objectives.

To achieve 'green' and affordable housing, there should be both an effective social housing policy and a strong concern for environment protection. In France, 'green' building policies emerged in the 1990s with the climate change debate, and the State has given building energy efficiency higher priority, to comply with the European community's strategy and action plan on renewable energy. In Israel, energy performance inspection and post-occupancy assessment practices could improve the energy efficiency of buildings, capitalizing on the new standards (SI 5281 and SI 5282). This could reduce housing operation costs, but application on a sufficient scale might involve massive State funding, incompatible with Treasury's neoliberal policy.

The SI 5281 standard tends to make apartments less affordable than they already are. Thus, the Forum 15 decision to adopt the standard as mandatory (see chapter "Introduction") may ber a strategy to raise prices. Local governments may favor gentrification, to gain higher income residents, while compelling poorer residents to relocate, due to their inability to pay increased rents or property taxes.

While the dilemmas raised by gentrification are well-known, 'green' gentrification is relatively new. The ecological functionality of neighborhoods has become an object of middle-class consumption, reflecting the interests, priorities and values of middle-class consumers. Through 'green' gentrification, the benefits of sustainability in terms of environmental goods – such as green space, and energy efficient buildings – concentrate in middle-class areas. This perpetuates socio-spatial inequality, worsening the ecological vulnerability of socially marginal groups, while facilitating their social exclusion. The tale of two countries presented here suggests that neither the mechanisms of the market nor the proclamations of a welfare state can easily overcome this dilemma. New thinking, can only emerge once the concept of "value" reflects not only the realities of the market, but also those of a planet which turns out to be distinctly limited in its resources.

Appendix 1

Municipality	Poverty rate 2015	Average income of the poorest (1st dec.) 2015	Average income of the richest (9th dec.) 2015	Ratio of standard of living (9th dec./1st dec.) 2015	Share of income taxes 2015	Median standard of living 2015
Paris	16.2	9864	63,781	6.5	−27.4	26,431
Boulogne-Billancourt	9.7	12,528	67,469	5.4	−27.5	31,268
Levallois-Perret	9.1	12,833	65,018	5.1	−26.8	30,504
Nanterre	21.2	9472	39,218	4.1	−17.5	19,170
Brétigny-sur-Orge	12.6	11,541	36,879	3.2	−17.7	21,875
Bordeaux	17.2	9868	44,568	4.5	−20.9	21,404
Tours	19.9	9469	36,447	3.8	−17.4	19,133
Angers	19.9	9576	36,538	3.8	−16.6	19,194
Mantes-la-Jolie	32.5	8147	29,218	3.6	−14.7	15,410
Reims	22.8	9205	35,312	3.8	−17.3	18,231
Lille	25.7	8710	37,099	4.3	−17.9	18,102
Lyon	14.6	10,720	44,743	4.2	−20.6	22,501
Mulhouse	33.2	8326	32,167	3.9	−15.9	15,371
Grenoble	19	9709	37,843	3.9	−17.6	20,066
Saint-Brieuc	19.8	9656	34,198	3.5	−16.4	18,838
Ivry-sur-Seine	27.9	7735	34,427	4.5	−16.7	17,406
Montpellier	26.5	8546	35,687	4.2	−17.6	17,640

Source: INSEE (2018)
Selected socio-economic data in municipalities which include at least one certified eco-quartier

© The Author(s), under exclusive license to Springer Nature Switzerland AG 2020
E. Machline et al., *Green Neighbourhoods and Eco-gentrification*,
SpringerBriefs in Environmental Science,
https://doi.org/10.1007/978-3-030-38036-6

Appendix 2: Post Occupancy Survey in Yavneh's 'Green' Neighborhood

The following data were collected among 187 residents that moved to the Neot Rabin 'green' neighborhood of Yavneh between 2012 and 2016. The majority (74%) settled in 2013–2015, and most (93%) purchased new apartments (5% bought them second-hand).

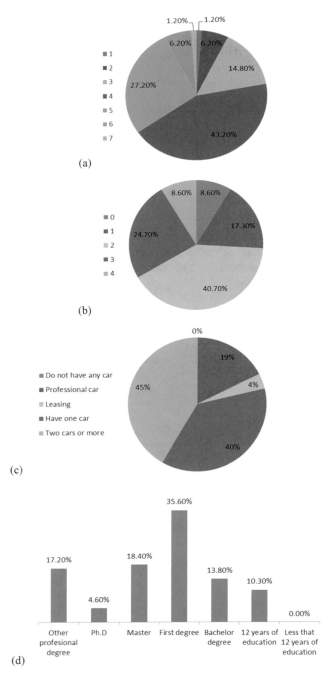

Socio-economic data for Neot Rabin. (**a**) Number of persons per households; (**b**) Number of children under 18 years old per household; (**c**) Educational level in the 'green' neighborhood; (**d**) Number of cars per household. (Source: original survey data, 2017)

References

Ahern, J. (2011). From *fail-safe* to *safe-to-fail*: Sustainability and resilience in the new urban world. *Landscape and Urban Planning, 100*(4), 341–343.

Alexander, E. R. (2012). Institutional design for value capture and a case: The Tel-Aviv metropolitan park. *International Planning Studies, 17*(2), 163–177.

Alvim, R., & Leite Lopes, J. S. (1990). Familles ouvrières, familles d'ouvrières. *Actes de la recherche en sciences sociales, 84*(1), 78–84.

Andrews, D., Sánchez, A. C., & Johansson, Å. (2011). Housing markets and structural policies in OECD countries.

Ariav Y., & Amir, M. (2011). Economic analysis and reexamination of Government Solar Energy Policy in Israel – Updated Status (in hebrew).

APUR (2007). Politique de la ville: observatoire des quartiers parisiens, rapport 2007 – La nouvelle géographie des quartiers prioritaires, Paris.

Atkinson, R., & Bridge, G. (2005). The new urban colonialism: gentrification in a global context.

Authier, R. J.-Y. (2007). Les "quartiers" qui font l'actualité. *Espaces et sociétés, 1*(128–129), 239–249.

Banzhaf, H. S., & McCormick, E. (2007). *Moving beyond cleanup: Identifying the crucibles of environmental gentrification. Andrew Young School of Policy Studies Research Paper Series, Working Paper 07–29, May.* Atlanta: Georgia State University.

Bar Ilan, Y., Pearlmutter, D., & Tal, A. (2010). *Building green: Promoting energy efficiency in Israel, Center for Urban and Regional Studies.* Haifa: Technion – Israel Institute of Technology.

Barker, K. (2004). Review of housing supply: delivering stability: securing our future housing needs: final report: recommendations. London: HM Treasury.

Bartlett, E., & Howard, N. (2000). Informing the decision makers on the cost and value of green building. *Building Research & Information, 28*(5–6), 315–324.

Baudin G. (2001). La mixité sociale : Une utopie urbaine et urbanistique, revue du CREHU, 12p.

Beeri, I., & Navot, D. (2013). Local Political Corruption: Potential structural malfunctions at the central–local, local–local and intra-local levels. *Public Management Review, 15*(5), 712–739.

Ben-Elia, N. (1998). *The municipal crisis in Israel, The management failure and the challenge of recovery [Hebrew].* Jerusalem: The Floersheimer Institute for Policy Studies.

Blanc, M. (2004). The changing role of the state in French housing policies: A roll-out without roll-back? *International Journal of Housing Policy, 4*(3), 283–302.

Blanc, M. (2010), The impact of social mix policies in France. *Housing Studies, 25*(2), 257–272.

© The Author(s), under exclusive license to Springer Nature Switzerland AG 2020
E. Machline et al., *Green Neighbourhoods and Eco-gentrification*,
SpringerBriefs in Environmental Science,
https://doi.org/10.1007/978-3-030-38036-6

Bond, L., Kearns, A., & Sautkina, E. (2012). Mixed evidence on mixed tenure effects: Findings from a systematic review of UK studies, 1995–2009. *Housing Studies, 27*(6), 748–782.

Boughriet, R. "Performance énergétique : Lancement du nouveau label Effinergie +" [archive], sur www.actu-environnement.com, Actu Environnement, 18 janvier 2012.

Boutaud, A. (2009). Les Agendas 21 locaux. Bilan et perspectives en Europe et en France, Lyon, Grand Lyon. Communauté urbaine Millénaire 3, 56 p.

Bradshaw, W., et al. (2005). *The costs and benefits of green affordable housing.* Cambridge (MA): New Ecology.

Bristol, A. (2005). Conclusions of ministerial informal on sustainable communities in Europe. Britain: UK Presidency.

Burton, E. (2000). The compact city: just or just compact? A preliminary analysis. *Urban studies, 37*(11), 1969–2006.

Calcalist. (2016). Calcalist Israeli newspaper. Demolation and reconstrcutction program in Neve Sharet-Tel Aviv. March 2016, retrieved from website: https://www.calcalist.co.il/local/articles/0,7340,L-3679298,00.html

Canelli, A. (2017). Acteurs et processus décisionnels lors des réhabilitations de friches industrielles – le cas de Hammarby Sjöstad. Recherche de bachelor en Géographie et environnement, Université de Genève.

Carmon, N. (2001). Housing policy in Israel: Review, evaluation and lessons. *Israel Affairs, 7*(4), 181–208.

Carmon, N., & Mannheim, B. (1979). Housing policy as a tool of social policy. *Social Forces, 58*(1), 336–351.

Catenaccio, P. (2009). Green advertising and corporate CSR communication: Hybriddiscourses, mutant genres. Retrieved 14 June 2011, from http://www.genresonthemove2009.unina.it/abstracts/Catenaccio%20Paola.pdf

CBS. (1981). CBS (Central Bureau of Statistics), Statistical abstract of Israel, Population socio-economic data No. 22.3, 1981.

Certification File. (2015). French Ministry of Housing and Sustainable development, Le referentiel ecoquartier 2016.

Central Bureau of Statistics. (2005). Israel Central Bureau of Statistics. Statistical Abstract of Israel 2005-No.61. 2005.

Charlot-Valdieu, C., & Outrequin, P. (2009). L'urbanisme durable: Concevoir un écoquartier.Ed. Le Moniteur. France.

Checker, M. (2011). Wiped out by the "Greenwave": Environmental gentrification and the paradoxical politics of urban sustainability. *City & Society, 23*, 210–229.

Chiu, R. (2002). Social equity in housing in the Hong Kong Special Administrative Region: A social sustainability perspective. *Sustainable Development, 10*, 155–162.

Clark, E. (2005). The order and simplicity of gentrification: A political challenge. In *Gentrification in a global context: The new urban colonialism* (pp. 261–269). Retrieved from: http://lup.lub.lu.se/record/620935

Clerval, A. (2010). Les dynamiques spatiales de la gentrification à Paris. *Cybergeo: European Journal of Geography.* Retrieved from http://cybergeo.revues.org/23231?lang=en&hc_location=ufi

Clerval, A., & Fleury, A. (2012). Urban policy and gentrification. A critical analysis using the case of Paris. Lidia Diappi. Emergent phenomena in housing markets. Gentrification, housing search, polarization. *Springer, 21*, 151–170.

Cohen, S., & Amir, T. (2007). From a public shelter to a protected space ("MAMAD"): The privatization of civil protection. In S. Cohen & T. Amir (Eds.), *Living forms, architecture and society in Israel* (pp. 127–143). Tel-Aviv: Xargol.

Cohen, N., & Margalit, T. (2015). There are really two cities here: fragmented urban citizenship in Tel Aviv. *International Journal of Urban and Regional Research, 39*(4), 666–686.

Cohen, C., Pearlmutter, D., & Schwartz, M. (2017). A game theory-based assessment of the implementation of green building in Israel. *Building and Environment, 125*, 122–128.

Comay, Y., & Kirschenbaum, A. (1973). The Israeli new town: An experiment at population redistribution. *Economic Development and Cultural Change, 22*(1), 124–134.

Commission for Environmental Cooperation. (2008). Opportunities and challenges- Secretariat report to Council Under Article 13 of the North American Agreement on Environmental Cooperation. Montreal.

Dampier, W. (1982). Ten years after Stockholm: A decade of environmental debate. *Ambio, 11*(4), 215–231.

Dargay, J., & Gately, D. (1997). Vehicle ownership to 2015: implications for energy use and emissions. *Energy Policy, 25*(14–15), 1121–1127.

de Chastenet, C., Belziti, D., Bessis, B., Faucheux, F., Le Sceller, T., Monaco, F. X., & Pech, P. (2016). The French eco neighbourhood evaluation model: Contributions to sustainable city making and to the evolution of urban practices. *Journal of Environmental Management, 176*, 69–78.

De Vaus, D. A. (2001). *Research design in social research*. London: Sage.

Decamps, A. (2011). La dynamique de la ségrégation urbaine à travers l'évolution des profils de quartiers: Étude sur l'agglomération bordelaise. *Revue d'Économie Régionale & Urbaine, 1*, 151–183.

Dery, D., & Schwartz-Milner, B. (1994). Who governs local government. *Tel Aviv: Hakibutz Hameuchad*.

Dikeç, M. (2007). *Badlands of the republic: Space, politics and urban policy*. Malden: Blackwell.

Donahue, S. (2004). How clean are green ads? Evaluating environmental advertising in contemporary media. In *Program in writing and rhetoric*. Stanford: Stanford University. Retrieved 14 June 2011, from http://www.stanford.edu/group/boothe/0405/PWR-Donahue.pdf

Donzelot, J., Mevel, C., & Wyvekens, A. (2003). *Faire société. La politique de la ville aux États-Unis et en France* (364 p). Paris: Seuil.

Dovman, Polina, Sigal Ribon, and Yossi Yakhin (2012): "The Housing Market in Israel 2008–2010: Are House Prices a 'Bubble'?" Israel Economic Review 10(1):1–38.

Efrat, E. (1984). *Urbanization in Israel*. London: Routledge.

Eichholtz, P., Kok, N., & Quiqley, J. (2010). *The economics of green building*. Berkeley: Maastrict University and University of California.

Elsinga, M., & Lind, H. (2013). The effect of EU-legislation on rental systems in Sweden and the Netherlands. *Housing Studies, 28*(7), 960–970.

Emelianoff, C. (2007). La ville durable: L'hypothèse d'un tournant urbanistique en Europe. *L'information géographique, 71*, 48–65.

Esping-Andersen, G. (1990). *The three worlds of welfare capitalism*. Cambridge, UK: Polity/Princeton: Princeton Univ. Press.

Flint, J., & Kearns, A. (2006). Housing, neighbourhood renewal and social capital: The case of registered social landlords in Scotland. *European Journal of Housing Policy, 6*(1), 3154.

Fuerst, F., & McAllister, P. (2011). Green noise or green value? Measuring the effects of environmental certification on office values. *Real estate economics, 39*(1), 45–69.

Gabay, H., Meir, I. A., Schwartz, M., & Werzberger, E. (2014). Cost-benefit analysis of green buildings: An Israeli office buildings case study. *Energy and buildings, 76*, 558–564.

Garde, A. (2009). Sustainable by Design? Insights From U.S. LEED-ND Pilot Projects. *Journal of the American Planning Association 75*(4), 424–40.

Gillespie, E. (2008). Stemming the tide of "greenwash". *Consumer Policy Review, 3*, 79–83.

Gonen, A. (2015). Widespread and diverse forms of gentrification in Israel. Gentrifications: Uneven development and displacement, 143–164.

Goulden, S. (2016). Constructing 'green building': Heterogeneous networks and the translation of sustainability into planning in Israel. In Actor Networks of Planning (pp. 27–43). Routledge.

Goulden, S., Erell, E., Garb, Y., & Pearlmutter, D. (2017). Green building standards as socio-technical actors in municipal environmental policy. *Building Research & Information, 45*(4), 414–425.

Gradus, Y. (1984). The emergence of regionalism in a centralized system: The case of Israel. *Environment and Planning D: Society and Space, 2*, 87–100.

Gradus, Y., & Einy, Y. (1981, May). Trends in core-periphery industrialization gaps in Israel. In Geographical Research Forum (Vol. 3, No. 1, pp. 25–37).

Gradus, Y., Razin, E., & Krakover, S. (2006). *The industrial geography of Israel*. New York: Routledge.

Hamedani, A. Z., & Huber, F. (2012). A comparative study of DGNB, LEED and BREEAM certificate systems in urban sustainability. The sustainable city VII: Urban regeneration and sustainability, 1121.

Hasson, S. (1981). Social and spatial conflicts: the settlement process in Israel during the 1950s and the 1960s. *L'espace Geographique 10*(3):169–179.

Haughton, G. (1999). Environmental justice and the sustainable city. In D. Satterthwaite (Ed.), *Sustainable cities* (pp. 79–62). London: Earthscan.

Holden, E., & Linnerud, K. (2007). The sustainable development area: Satisfying basic needs and safeguarding ecological sustainability. *Sustainable Development, 15*, 174–185.

Hopwood, B., Mellor, M., & O'Brien, G. (2005). Sustainable development: Mapping different approaches. *Sustainable Development, 13*, 38–52.

INSEE (Institut National de la Statistique et des Études Économiques). (2010). Commune: Paris (75056) – Thème: Évolution et structure de la population. Paris: INSEE.

INSEE (Institut National de la Statistique et des Études Économiques). (2018). National census at census block level. Paris: INSEE.

Israeli Builders Association. (2015). Review of Developments in 2014 and Forecast for 2016-2015, Construction and infrastructure department.

Israel Water Authority. (2009). Policy Issues – Water and Energy. Planning Division.

Israeli Ministry of the Interior and Ministry of Housing. (1996). Socio-Economic Ranking of Local Governments in Israel. Government Printers, Jerusalem.

Israeli Tax Authorities. (2017). Israeli Ministry of Finance, Israeli Tax authority website, Real Estate information, retrieved from website: https://taxes.gov.il/

Jain, A. K. (2001). Corruption: A review. *Journal of economic surveys, 15*(1), 71–121.

Kahn, M. E., & Kok, N. (2012). The value of green labels in the California housing market: An economic analysis of the impact of green labeling on the sales price of a home. Report, July.

Kats, G. (2003). *The costs and financial benefits of green buildings—A report to California's sustainable building task force*. Washington, DC: U.S. Green Building Council.

Karahan, G. R., Coats, R. M., & Shughart, W. F. (2006). Corrupt political jurisdictions and voter participation. *Public Choice, 126*(1–2), 87–106.

Kaufman, B. J. (2010). Green Homes Outselling the Rest of the Market. Seattle Daily Journal of Commerce, February 18. http://www.djc.com/news/en/12015059.html

Kearns, A., & Mason, P. (2007). Mixed tenure communities and neighborhood quality. *Housing Studies, 22*(5), 661–691.

Keil, R. (2005). Progress report—urban political ecology. *Urban Geography, 26*(7), 640–651.

Kibert, C. J. (2013). *Sustainable construction: Green building design and delivery* (3rd ed.). Hoboken: John Wiley & Sons.

Kipnis, B. (1987). Geopolitical ideologies and regional strategies in Israel. *Tijdchrift voor Economische en Sociale Geografie, 78*, 125–138.

Kirszbaum, T. (2008). *Mixité sociale dans l'habitat: Revue de la littérature dans une perspective comparative* (141 p). Paris: La Documentation française.

Kot, H., & Katz, D. (2013). Green Building Costs of Residential Buildings in Israel, The Israeli Green building Council.

Koester, V. (1980). Nordic countries' legislation on the environment with special emphasis on conservation: a survey (No. 14). International Union for Conservation of Nature and Natural Resources.

Krueger, R., & Agyeman, J. (2005). Sustainability schizophrenia or "actually existing sustainabilities?" Toward a broader understanding of the politics and promise of local sustainability in the US. *Geoforum, 36*(4), 410–417.

Krueger, R., & Savage, L. (2007). City-regions and social reproduction: a 'place'for sustainable development?. *International Journal of Urban and Regional Research, 31*(1), 215–223.

Lacroix, V., & Zaccaï, E. (2010). Quarante ans de politique environnementale en France : Évolutions, avancées, constante. *Revue française d'administration publique 2, 134*, 205–232.

Larsen, H. G., & Hansen, A. L. (2008). Gentrification—gentle or traumatic? Urban renewal policies and socioeconomic transformations in Copenhagen. *Urban Studies, 45*(12), 2429–2448.

Lautre, Y. (2018). Chômage, précarités, pauvreté en France: Archives 2011–2018.

LEED ND Guidelines (2016). U.S. Green Building Council. Certification. [En ligne] http://www.usgbc.org/certification, 2015.

Lees, L. (2000). A reappraisal of gentrification: towards a 'geography of gentrification'. *Progress in human geography, 24*(3), 389–408.

Lees, L. (2003). Super-gentrification: The case of Brooklyn heights, New York city. *Urban studies, 40*(12), 2487–2509.

Lefebvre, B., Mouillart, M., & Occhipinti, S. (1991). *Politique du logement: 50 ans pour un échec*. Paris: L'Harmattan.

Lele, S. M. (1991). Sustainable development: A critical review. *World Development, 19*(6), 607–621.

Lelevrier C. (2006). Les mixités sociales. *Problèmes politiques et sociaux, 29*.

Lévy-Vroelant, C. (2007). Urban Renewal in France: What or who is at stake?. *Innovation, 20*(2), 109–118.

Loison, M. (2007). The implementation of an enforceable right to housing in France. *European Journal of Homelessness, 1*, 185–197.

Machline, E., Pearlmutter, D., & Schwartz, M. (2018). Parisian eco-districts: low energy and affordable housing?. *Building Research & Information, 46*(6), 636–652.

Markets Department (2000). Annual Report on Exchange Arrangements and Exchange Restrictions, 2000. International Monetary Fund.

Marom, N. (2014). Planning as a Principle of Vision and Division: A Bourdieusian View of Tel Aviv's Urban Development, 1920s—1950s. *Environment and Planning A, 46*(8), 1908–1926.

Massimo, D. (2012). Emerging issues in real estate appraisal: market premium for building sustainability. *Aestimum*, 653–673.

Milken Institute. (2015). Milken Institute Israel Center. Toward affordable housing in Israel, [En ligne] https://milkeninnovationcenter.org/wp-content/uploads/2015/10/Affordable-Housing-ENG.pdf, 2014.

MoNI. (2010). The Israeli Ministry of National Infrastructures, Natural Gas Authority, "The Development of the Israeli Gas Sector," February 2010.

National Economic Council. (2014). Future Housing Needs of the Israeli Population, Head Government Office.

Observatory for Building Degradation Prevention in Paris (2012): Observatoire de la prévention de la dégradation des immeubles anciens à Paris. Les chiffres du logement social à Paris en 2011 – Edition 2012, Paris.

OECD (2013). OECD – Health Policy and Data: Health Division Website. Retrieved from http://www.oecd.org/department/0,2688,en_2649_33929_1_1_1_1,00.html

OECD. (2014). OECD Factbook 2014: Economic, Environmental and Social Statistics, OECD Publishing, Paris, https://doi.org/10.1787/factbook-2014-en

Olander, Y. (1999). A "Green" Neighborhood in the City of Kfar Saba, Israel: An Applicational Model. (Master Plan KS/1/60 Israel).

Paz-Frankel, E. (2012). Home prices double OECD average in salary terms. Globes.

Pierson, J. (2002). *Tackling social exclusion*. London: Routledge.

Prime Minister's Office (2008). Socio-Economic Agenda for Israel, 2008-2010. Jerusalem: Prime Minister's Office. National Economic Council. http://www.pmo.gov.il/PMO/PM+Office/Departments/econ20082010.htm

Rabinovich, D. (2007). A private apartment, an apartment building, a public space: The Israeli housing unit between private space and common property. In S. Cohen & T. Amir (Eds.), *Living forms, architecture and society in Israel* (pp. 144–163). Tel-Aviv: Xargol.

Razin, E. (2004). Needs and impediments for local government reform: Lessons from Israel. *Journal of Urban Affairs, 26*(5), 623–640.

Recast, EPBD. (2010). Directive 2010/31/EU of the European Parliament and of the Council of 19 May 2010 on the energy performance of buildings (recast). Official Journal of the European Union, 18(06), 2010.

Sampson, R. J., Raudenbush, S. W., & Earls, F. (1997). Neighborhoods and violent crime: A multilevel study of collective efficacy. *Science, 277*(5328), 918–924.

Sanchez, M. (2018). Les relations entre les mesures écologiques et les mesures sociales dans l'écoquartier de Vesterbro à Copenhague, In : Pech P. Écoquartiers et biodiversité. Quel rôle social joue la biodiversité dans les écoquartiers ? Paris : Éditions Johanet.

Shadar, H. (2004). Between east and west: Immigrants, critical regionalism and public housing. *The Journal of Architecture, 9*(1), 23–48.

Sharifi, A., & Murayama, A. (2013). A critical review of seven selected neighborhood sustainability assessment tools. *Environmental impact assessment review, 38*, 73–87.

Silberberg, R. (1973). Population Distribution in Israel, 1948–1972, The Economic Planning Authority, Jerusalem (in Hebrew).

Simon, P. (1995). La politique de la ville contre la ségrégation. Ou l'idéal d'une ville sans divisions. *Les Annales de la recherche urbaine, 68–69*, 27–33.

Slater, T. (2004). Municipally managed gentrification in south Parkdale, Toronto. *Canadian Geographer/Le Géographe Canadien, 48*(3), 303–325.

Smith, N. (1996). The New Urban Frontier: Gentrification and the Revanchist City (Routledge, London).

Souami, T. (2009). *Écoquartiers, secrets de fabrication. Analyse critique d'exemples européens.* Éditions les Carnets de l'info. Paris, Coll. Modes de ville, 208 p. ISBN: 978-2-9166-2844-8.

Statistics Sweden. (2018). Average monthly salary, gender and year [Dataset 2018].

Sverdlov, A., & Dolev, S. (2009). Handling peak demand for electricity in Israel: Analysis of the problem and offer solutions to policy. In Tel Aviv, Israel: Israel Energy Forum.

Szibbo, N. (2016). Assessing neighborhood livability: Evidence from LEED® for neighborhood development and new urbanist communities. *Articulo – Journal of Urban Research* [Online].

Theys, J. (2002). L'approche territoriale du "développement durable", condition d'une prise en compte de sa dimension sociale. Développement durable et territoires. *Économie, géographie, politique, droit, sociologie*, (Dossier 1).

Trajtenberg, M. (2012). Creating a more just Israeli society. Official summary of the Trajtenberg Report.

Tunkrova, L. (2008). The Nordic Countries' 'Exceptionalism' in EU Environmental Policy. Contemporary European Studies, (02), 21–46.

Tunstall, R., & Fenton, A. (2006). In the mix, a review of mixed income, mixed tenure and mixed communities: what do we know. York, Housing Corporation, Joseph Rowntree Foundation & English Partnerships.

Tuot, T. (2007). Le Grenelle de l'Environnement: Rapport Général, Ministère de l'écologie, du développement et de l'aménagement durables, Paris.

Urban Solidarity and Renewal Act. (2000). SRU (2000) Loi Relative à la Solidarité et au Renouvellement Urbains, 14 December. Available at http://www.legifrance.gouv.fr/affich-Texte.do?

USGBC, LEED (2011). US Green Building Council. Inc.,. International Monetary Fund 2000: International Monetary Fund. Monetary, & Capital.

U.S. Green Building Council. (2016). U.S. Green Building Council. Certification. [En ligne] http://www.usgbc.org/certification, 2015.

Vine, E., Hanrin, J., Eyre, N., Crossley, D., Maloney, M., & Watt, G. (2003). Public policy analysis of energy efficiency and load management in changing electricity businesses. *Energy Policy, 31*, 405–430.

WCED. (1987). Our common future—The Brundtland report. World Commission on Environment an Development. Oxford University Press, Oxford.

Wiley, J., Benefield, J., & Johnson, K. (2010). Green design and the market for commercial office space. *The Journal of Real Estate Finance and Economics, 41*(2), 228–243.

Williams, K., Burton, E., & Jenks, M. (Eds.). (2000). *Achieving sustainable urban form.* London: Spon.

Yehezkel, Y. (2008). Retailers' choice of product variety and exclusive dealing under asymmetric information. *The RAND Journal of Economics, 39*(1), 115–143.

Index

A

Affordable housing, 12, 14–16, 37, 38, 46, 75
    development, 15
    location, 42, 43
    in Tel Aviv, 45
Apartment owners, 36
Apartment prices and rents, 89
Apartment sales price, 50
Apartment size, 51, 52
Area median income (AMI), 44

B

Bretigny, 81
British rating system, 3

C

China Development Bank, 66
Cities for Climate Protection Initiative, 37
"Cités ouvrières", 7
City councils, 5
Claude Bernard eco-quartier, 76, 78
Clause Bois Badeau eco-quartier, 79, 80
Computer energy simulation, 53
Covenant of Mayors, 71

D

Demographic Growth Unit, 59
Diagnostic de performance Energetique
        (DPE), 70
Dimona, 47, 55
    municipality, 56

residents, 59
role, 55
Shahar neighborhood, 58, 59
solar energy transformation, 55
stakeholders, 59

E

Eco-districts, 11, 54, 62, 64
Eco-gentrification, 17, 19, 20
Economy, 27
Eco-quartier case studies, 74
Eco-quartier distribution, 75
Eco-quartier label, 66
Eco-quartiers, 19, 69, 76–79, 95
Energy Performance of Buildings Directive
        (EPBD), 70
EnergyUI software, 53
Environmental design approach, 64
Esping-Andersen's classification, 6
Europe
    cost-price principle, 63
    dwellings, 63
    environmental aspects, 61
    social equity, 62
    social justice, 61

F

France
    authorities, 69
    case studies, 2, 69–85
    eco-districts, 71–73
    EPC, 70

France (*cont.*)
  GHG reduction, 72
  green value, 70
  municipalities, 74
  national urban renewal programme, 72
  OECD, 74
  social mix, 73
French Eco-quartiers social diversity, 14
French housing policy, 7
French policy tools, 92
Frequel-Fontarabie eco-quartier, 77

G
Gentrification effect, 12, 15, 16, 90
German Colony, 16
GovMap, 22
Green apartment, 26, 28
Green building, 1, 49
  certification, 89
  construction, 2
  design and construction, 3
  economic and social implications, 2
  in France, 10–12
  implementation, 3
  Israeli cases, 2
  local governments, 5
  organizations, 3
  Scandinavian countries, 2
  upper middle-class dweller, 2
Green building development
  energy efficiency, 9
  ISI, 8
  MoNI, 9
  thermal insulation, 8
Green Building Standard, 9
Green' certified buildings, 19
Green construction cost, 14
Green' neighborhood location, 47, 48, 89
Green Park master plan, 33
Green Park neighborhood, 35
Green Park project, 32, 35, 36, 45
Green rating systems, 1
Green value, 14, 24, 88
Greenhouse gas (GHG) emissions, 3, 71, 88
Greentrification, 92

H
Habitations à loyer modéré (HLM's), 8
Hammarby Sjöstad's project, 65
Head of the Engineering Department, 37, 43, 49
High Energy Performance (HPE), 10

Housing construction, 4
Housing Division, 6
Housing policies, 5–8
  market distortions, 5
  nation-building and developing, 6
  public construction, 6
Housing project, 47

I
Israel
  air pollution, 27
  apartment, 26
  case studies, 28–29
  construction costs, 21, 27
  consumption, water and electricity, 27
  costs, 21
  economic research, 21
  economy, 27
  evacuation and reconstruction, 30, 32
  green neighborhoods, 28
  green premium, 22, 23
  household, 26
  housing market, 25
  housing prices, 24, 28
  housing units, 24
  individual buildings, 22
  pollution costs, 27
  population, 27
  real estate prices, 21
  urban localities, 22
  urban renewal, 29
Israeli Builders Association, 23, 26
Israeli Central Bureau of Statistics, 31
Israeli governance regime, 4–5
Israeli Government, 6
Israeli Green Building Council (ILGBC), 9
Israeli housing prices, 90
Israeli Lands Authority, 58
Israeli local authorities, 4
Israeli Ministry of Environmental
    Protection, 26
Israeli residential building field, 6
Israel Ministry of Environmental
    Protection, 8
Israel power, 5
Israel's building industry, 6
Israel's Central Bureau of Statistics, 46
Israel Standards Institute (ISI), 8
Israel Tax Authority website, 22

J
Jewish immigration, 6

K
Kyoto Protocol, 10

L
Labor hegemony, 6
Law of Public housing, 88
Leadership in energy and environmental
    design (LEED), 3
    guidelines, 44
Leadership in energy and environmental
    design for neighborhood development
    (LEED ND), 38, 39
Local governments, 5
Low Consumption Building, 10

M
Ministry of Housing, 29
Ministry of National Infrastructures
    (MoNI), 8
Mixed Income Diverse Communities, 43

N
Negev development towns, 55
Neighborhood effects, 73
Neighborhood scale, 11
Neot Rabin 'green' neighborhood, 52
Neot Rabin residents, 54
Neve Sharett project, 46
Non-profit housing, 63

P
Paris population, 75
Pays de France Eco-quartier, 85
Pinui-binui project, 30, 45

R
Real estate prices, 2, 17, 35, 50, 89
Redevelopment process, 34
Reims, 82
Renovation plan, 64
Residents Council, 53
Rotem Industrial Complex, 55

S
Sede Dov Airport, 38
Shahar neighborhood project, 58
Social diversity, 2, 15
Social equity, 19
Social housing distribution, 83
"Social model", 32
Social polarization, 47
Social welfare policies, 1
Socio-economic data, 51
Solar water heating systems, 36
Standards Institute of Israel (SII), 21–22
Substantial political decentralization, 4
Sustainability, 18
Sustainable development, 18

T
TA 3700 neighborhood, 42
TA 3700 project, 37, 44
TAMA 38, 30
Tel Aviv 3700 project, 37, 41, 45
Tel Aviv metropolitan area, 22, 47
Tel Aviv municipality, 44
2012 Thermal Regulation, 11

U
Urban development project, 91
Urban renewal authorities, 45
Urban Solidarity and Renewal Act, 14

V
Vesterbro
    rehabilitation, 64
    unemployment, 64
    urbanization, 64

W
World War II, 7

Z
Zero-default city, 17
Zones d'Aménagement Concerté (ZAC), 8